Lecture Notes in Mathematics
A collection of informal reports and seminars
Edited by A. Dold, Heidelberg and B. Eckmann, Zürich

129

K. H. Hofmann
Tulane University, New Orleans, LA/USA

The Duality of Compact Semigroups and C*-Bigebras

Springer-Verlag
Berlin · Heidelberg · New York 1970

Title No. 3286

We propose a complete duality theory for the category of compact Hausdorff topological semigroups in terms of commutative C*-bigebras with identity. A bigebra (formerly called hyperalgebra or Hopf algebra) is an algebra with a comultiplication. In a more expository introductory part, we discuss the concept of a bigebra over a category, which plays an important role in this context. The second, more technical part starts with a discussion of C*-algebras and their tensor products. Finally, after the establishment of the duality, we show how it generalizes the Tannaka duality theory for compact groups, and the Pontryagin duality for compact abelian groups.

0. Introduction

One of the best known duality theories is the one between compact abelian groups and discrete abelian groups, discovered by PONTRYAGIN in the early thirties. A duality theory for not necessarily commutative compact groups was later described by TANNAKA, and its most modern forms are comparatively recent and go back to HOCHSCHILD [8].

All of these duality theories are closely tied up with the representation theory of these groups. At first sight this observation makes the outlook to possible generalizations for wider classes of compact objects (compact semigroups, say) seem gloomy. It is indeed well known and has been accepted with pessimistic fatalism that no useful representation theory is available for the category of all compact semigroups. The reason of this default is the absence in general compact semigroups of an invariant integral whose support is the entire semigroup. (In fact the only compact semigroups having a two sided invariant integral with full support are the groups.)

On the other hand, if one proceeds to the end of the spectrum of compact objects, namely those objects having no additional structure whatsoever, then again we have a completely satisfactory duality theory: namely, the duality between compact spaces and commutative C*-algebras which is based on the theorem of GELFAND and NAIMARK. It is therefore natural to look for dual objects of compact semigroups in the class of commutative C*-algebras with an additional element of structure. And indeed we will be able to establish a complete duality between the category of compact semigroups and the category of C*-algebras with a comultiplication. Algebras of certain types which, in addition, are endowed with a comultiplication have traditionally been called hyperalgebras by the algebraic geometers [2] and Hopf algebras by the algebraic topologists. HOPF observed there occurrence first in the cohomology ring of compact Lie groups (we will briefly outline this example in our discourse). Authors like DIEUDONNÉ and CARTIER (see e.g. [2]) used hyperalgebras extensively in the theory of formal Lie groups and algebras. Yet we will follow the more recent terminology which is sanctified by BOURBAKI. He calls an object without a multiplication but with a comultiplication a cogebra, and an object with a multiplication and a comultiplication a bigebra. Since we operate on C*-algebras we call a C*-algebra with a comultiplication in most cases a C*-bigebra.

The first part of our discussion is of a more narrative nature and describes the categorical background which we will make quite concrete by pointing out many relevant examples.

The brief Section 1 is devoted to a definition of duality, Section 2 to the idea of a multiplicative category.

In Sections 3, 4, 5 we discuss the concept of a bigebra in the spirit of category theory by starting with algebras, then proceeding to cogebras (the dual concept, and then finally by synthesizing the two to the concept of a bigebra. This part of the discussion is enriched by examples in order that the reader who considers a bigebra an unfamiliar if not useless object may get accustomed to the idea that bigebras in general are of fundamental importance in many different areas.

Since in Part II we apply these ideas to the category of C*-algebras, we have to devote Section 6 to some aspects of their structure theory, although basically we take the attitude that the reader is familiar with or willing to check the more elementary facts of this theory in the literature, e.g. in Dixmier's book about this subject [4].

What is needed specifically in our context is a tensor product for C*-algebras, without which the concept of a bigebra could not even be formulated. Even though there is a considerable literature about tensor products of C*-algebras, unfortunately not all questions seem to be clarified. Fortunately for our main applications, we are on excellent ground, since in the case of abelian C*-algebras the tensor product is perfectly understood. Still we discuss the C*-tensor product in somewhat greater generality in Section 7.

This then enables us to go directly from there to the introduction of C*-bigebras and a discussion of basic properties which takes place in Section 8.

In the very short Section 9 we point out the relation between coproducts and tensor products of C*-algebras.

We are then prepared for the duality theory of compact semigroups which now follows effortless and can be recognized as being in many respects a special case of a categorical pattern of great generality. Nevertheless the exploitation and technical details of the present theory are not of a categorical nature but require insight into the special structures at hand: that is understanding of the structure of compact semigroups as of C*-algebras as well. In Section 10 we describe this duality and work out examples of translating properties of compact semigroups into properties of C*-bigebras and vice versa. For instance, the well known fact that any compact semigroup has a unique minimal ideal will appear as a consequence of a much more general theorem about C*-bigebras, namely that every C*-bigebra with identity is colocal, and what this means will be explained in the text. Some quick applications of the duality follow.

The duality between the category of compact semigroups and certain C*-bigebras associates with a compact semigroup again an object which has both an algebraic and a topological structure, although in the case of C*-algebras one has the significant feature that they are "much more algebraic" than other classes of Banach algebras, since topological properties frequently are implied by the algebraic ones. Yet in the spirit of the classical duality theories one would wish to assign a purely algebraic object to a compact semigroup as a dual object. In view of what was said earlier the possible success of such an attempt is questionable; however, in Section 12 we explore to what extent this program is still feasible. In particular we associate with each C*-bigebra an (algebraic) involutive bigebra in a functorial fashion. The original C*-bigebra cannot always be recovered from it, and the extent to which this is possible gives a first indication of the degree of success in finding purely algebraic category as a dual if not for all so at least for a wide class of compact semigroups--including groups. It turns out, perhaps not unexpectedly, that the category of compact semigroups which is well behaved in this respect is the category of semigroups which have sufficiently many linear representations on finite dimensional vector spaces. Such semigroups will be called Peter-Weyl semigroups (for obvious reasons indicating the relationship to the theorem of Peter and Weyl for compact groups).

In Section 13 we derive the classical duality theories from our general one. Following the ideas outlined in HOCHSCHILD's book about Lie groups [8] we add some new facets to shed more light on the question why the existence of Haar measure distinguishes compact groups so significantly from compact semigroups with respect to duality and representation theory. The abstract form of the integral in bigebras has recently been utilized by SWEEDLER [16].

In Section 14 we derive new duality theorems expressing the dual category of the category of totally disconnected compact semigroups with identity in terms of biregular bigebras over the complex numbers. For groups and semilattices one obtains particularly smooth results. These duality theorems in some respect are more algebraic than the general duality in terms of C*-bigebras and are more intrinsic than the general Tannaka duality which involves the somewhat extraneous device of a gauge function.

Section 15, finally, is devoted to the Pontryagin duality of compact and discrete abelian groups in terms of our general duality theory. The first result (15.5), still quite general, shows how the semigroup bigebra of a discrete involutive semigroup fits into our scheme. All groups are involutive relative to inversion, and all abelian semigroups are involutive in a trivial fashion. The variety of objects functorially associated with a discrete involutive semigroup is illustrated in the simple example of the infinite cyclic semigroup, whose dual (in a sense) turns out to be the complex unit disc under multiplication. We discuss the deficiencies of this duality and proceed to show in what respect the situation is drastically better in the case of discrete groups and, perhaps somewhat surprisingly, semilattices. In fact the Pontryagin duality between compact abelian groups and discrete abelian groups has a complete analogue in the duality between compact totally disconnected semilattices with identity and discrete semilattices with identity. We will not, at present, carry out the duality theory, which would cover both of these as special cases, namely the duality between compact abelian Clifford semigroups with identity and totally disconnected semilattice of idempotents and discrete abelian Clifford semigroups with identity which from a somewhat different viewpoint was considered by Schneperman.

A first exposition of the present theory was presented by the author in a course about Compact Groups at Tulane University in 1966/67 while he was a Fellow of the Alfred P. Sloan Foundation. In a sequence of three lectures delivered at the Conference about Topological Semigroups at the University of Florida in Gainesville in April 1969 he reported about some of these developments. Naturally, the topic drew considerable profit from many fruitful discussions with P. S. Mostert. The support by the Alfred P. Sloan Foundation and the National Science Foundation is gratefully acknowledged. Thanks go to Connie Carrier for the care with which she prepared this typescript.

Table of contents

Part I

Part II

Section 12. Involutive bigebras

Section 13. Dualities for compact groups

Section 14. Dualities for totally disconnected compact semigroups

Section 15. Discrete involutive semigroups and duality

PART I

1. Definition of Duality

We presuppose familiarity with the elementary concepts of category theory such as category, functor, cofunctor (i.e. contravariant functor), object, morphism, natural transformation etc.

1.1. DEFINITION. Let two categories $\mathfrak{A}$, $\mathfrak{B}$ be given, and suppose that $S: \mathfrak{A} \longrightarrow \mathfrak{B}$ and $T: \mathfrak{B} \longrightarrow \mathfrak{A}$ are both functors or both cofunctors. If $ST: \mathfrak{B} \longrightarrow \mathfrak{B}$ is naturally isomorphic to the identity functor of $\mathfrak{B}$ and $TS: \mathfrak{A} \longrightarrow \mathfrak{A}$ is naturally isomorphic to the identity functor of $\mathfrak{A}$, then the categories $\mathfrak{A}$ and $\mathfrak{B}$ are called <u>equivalent</u>, if both S and T are functors, and <u>dual</u>, if both S and T are cofunctors. In the first case the pair (S,T) is called an <u>equivalence</u> between $\mathfrak{A}$ and $\mathfrak{B}$, in the second it is called a <u>duality</u> between $\mathfrak{A}$ and $\mathfrak{B}$.

1.2. Let $\mathfrak{A}$ be any category, and let $\mathfrak{A}^{\circ}$ be the same class of morphisms but with a new composition which is defined exactly for those pairs $(f,g) \in \mathfrak{A} \times \mathfrak{A}$ for which gf in $\mathfrak{A}$ is defined, and the composition is then given by $f * g = gf$. Then $\mathfrak{A}^{\circ}$ is a category. It is the category "with the same objects as $\mathfrak{A}$, but with all arrows reversed". The identity function $I: \mathfrak{A} \longrightarrow \mathfrak{A}^{\circ}$ then is a cofunctor, and (I,I) is a duality between $\mathfrak{A}$ and $\mathfrak{A}^{\circ}$. We call $\mathfrak{A}^{\circ}$ <u>the opposite category</u> of $\mathfrak{A}$.

The device of the opposite category (also called the dual category by some writers) makes the distinction between duality and isomorphy superfluous from a formal point of view, for the categories $\mathfrak{A}$ and $\mathfrak{B}$ are dual if and only if the categories $\mathfrak{A}^{\circ}$ and $\mathfrak{B}$ are equivalent. However, this is not how one operates in practice. The duality between $\mathfrak{A}$ and $\mathfrak{A}^{\circ}$ is a perfectly useless one from the practical point of view. The statement that the dual category of the category of compact abelian groups is the category whose objects are compact abelian groups "in which all arrows point the other way" does not give any new insights. Yet the observation that the category in question is dual to the category of discrete abelian groups (under a specifically given duality) is in fact the full extent of the structure theory of compact abelian groups. It is for this reason that we maintain the concept of duality separately from the concept of isomorphy of categories and that we place little if any emphasis on the opposite category of a category.

1.3. DEFINITION. Let $\mathfrak{A}'$ be a subcategory of $\mathfrak{A}$ and $\mathfrak{B}'$ a subcategory of $\mathfrak{B}$. Let $I:\mathfrak{A}' \longrightarrow \mathfrak{A}$, $J:\mathfrak{B}' \longrightarrow \mathfrak{B}$ be the inclusion functors. If (S',T') is a duality between $\mathfrak{A}'$ and $\mathfrak{B}'$ and (S,T) one between $\mathfrak{A}$ and $\mathfrak{B}$, we say that (S',T') and (S,T) are compatible if $JS' \cong SI$ and $IT' \cong TJ$.

2. Multiplicative categories

Some aspects of bigebras are most economically discussed in categorical language. The proper setting is that of a multiplicative category.

2.1. DEFINITION. A multiplicative category is a pair $(\mathfrak{A},\otimes)$ consisting of a category $\mathfrak{A}$ and an associative and commutative functor $\otimes : \mathfrak{A}\times\mathfrak{A} \longrightarrow \mathfrak{A}$.

Specifically, this means the following:

(a) The functor diagram

$$\begin{array}{ccc} \mathfrak{A}\times\mathfrak{A}\times\mathfrak{A} & \xrightarrow{\otimes\times I_{\mathfrak{A}}} & \mathfrak{A}\times\mathfrak{A} \\ {\scriptstyle I_{\mathfrak{A}}\times\otimes}\downarrow & & \downarrow{\scriptstyle\otimes} \\ \mathfrak{A}\times\mathfrak{A} & \xrightarrow{\otimes} & \mathfrak{A} \end{array}$$

is commutative.

(b) The functor diagram

$$\begin{array}{ccc} \mathfrak{A}\times\mathfrak{A} & \xrightarrow{T} & \mathfrak{A}\times\mathfrak{A} \\ & {\scriptstyle\otimes}\searrow \quad \swarrow{\scriptstyle\otimes} & \\ & \mathfrak{A} & \end{array}$$

is commutative where $T:\mathfrak{A}\times\mathfrak{A} \longrightarrow \mathfrak{A}\times\mathfrak{A}$ is the functor defined by $T(f,g) = (g,f)$.

We say that $(\mathfrak{A},\otimes)$ has an identity E if E is an object of $\mathfrak{A}$ such that there are natural isomorphisms $E\otimes A \cong A \cong A\otimes E$ for all objects A of $\mathfrak{A}$.

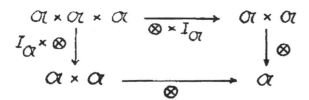

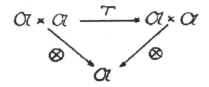

Note that any two identities of $\mathfrak{A}$ are isomorphic.

It may be well to remember at this point that "equality" of functors (which has to be agreed upon when the talk is about commutative functor diagrams) means their natural isomorphy.

Examples of multiplicative categories appear everywhere:

2.2. If a category $\mathfrak{A}$ has finite products (resp., coproducts), then $(\mathfrak{A},\times)$ (resp., $(\mathfrak{A},\oplus)$) are multiplicative categories. This applies particularly to the categories of sets, topological spaces, compact topological spaces, abelian groups, R-modules, etc.

One special case perhaps deserves to be pointed out specifically:

2.2.a. Let $\mathcal{R}_a$ be the category of commutative rings with identity. Then this category has finite coproducts (in fact it has arbitrary colimits); specifically for a two commutative rings A and B with identity the diagram $A \longrightarrow A\otimes B \longleftarrow B$ with the natural maps (such as $a \longrightarrow a\otimes 1$) is a coproduct, where $\otimes$ is the ordinary tensor product over $\mathbb{Z}$ and where $A\otimes B$ has the familiar ring structure which in fact may be imposed on the tensor product of any two rings (commutative or not, with or without identity). The category $(\mathfrak{A},\otimes)$ is then a multiplicative category.

Let us note that this prevails for the category of arbitrary rings with identity, but the tensor product is no longer the coproduct. Thus on the category $\mathcal{R}$ of rings with identity there are two different multiplicative structures $(\mathcal{R},\otimes)$ and $(\mathcal{R},\amalg)$, where $\amalg$ is the coproduct in the category $\mathcal{R}$.

2.3. Let R be a ring with identity (such as the ring $\mathbb{Z}$ of integers). Let $\otimes_R$ denote the tensor product of R-modules, and $\mathcal{G}^R$ the category of R-modules. Then $(\mathcal{G}^R,\otimes_R)$ is a multiplicative category.

Example 2.3 in fact is the classical case motivating the concept of a multiplication in a category. There are ramifications of the classical concept which are worth mentioning:

2.3.a. Let Grad $\mathcal{G}^R$ be the category of graded R-modules and $\otimes_R^*$ be the tensor product of graded R-modules (defined by $(A \otimes_R^* B)^i = \oplus\{A^p \otimes_R B^q : p+q = i\}$); then (Grad $\mathcal{G}^R,\otimes_R^*$) is a multiplicative category.

There is no difficulty in abstracting from this example the concept of a category of graded objects over a multiplicative category with finite coproducts and creating a multiplication on it in the very same fashion.

Tensor products themselves (in a more narrow sense) are in fact quite prevalent. They arise naturally in the solution of the universal problem of classifying bilinear maps.

While the general existence and uniqueness of tensor products does not present any difficulties in view of modern categorical methods such as the coadjoint functor existence theorem, the concrete description may sometimes be somewhat delicate. The following example is a point in case; we will rely on this example at a later point.

2.4. Let E and F be real or complex Banach spaces. The algebraic tensor product $E \otimes F$ can be given a norm which for an element $z \in E \otimes F$ is defined as follows: Let $A(z)$ be the set of all finite sequences $((x_1,y_1),\ldots,(x_k,y_k))$, $x_i \in E$, $y_i \in F$ such that $z = x_1 \otimes y_1 + \ldots + x_k \otimes y_k$. Then $\|z\| = \inf\{ \|x_1\| \, \|y_1\| + \ldots + \|x_k\| \, \|y_k\| : ((x_1,y_1),\ldots, (x_k,y_k)) \in A(z)\}$ is indeed a norm on $E \otimes F$. The completion of the normed space so arising is defined, denoted by $E \hat{\otimes} F$ and is called _the projective tensor product of the Banach spaces E and F_. It has the desired universal property: In fact any continuous bilinear function $f: E \times F \longrightarrow G$ into a Banach space G determines a unique continuous linear map $f': E \hat{\otimes} F \longrightarrow G$ such that $f(x,y) = f'(x \otimes y)$, where we naturally consider the algebraic tensor product $E \otimes F$ as a subvector space of $E \hat{\otimes} F$. The category of Banach spaces and continuous linear maps together with the projective tensor product forms a multiplicative category.

It may be observed, that the projective tensor product may be constructed for locally convex spaces in place of Banach spaces. For details we refer to GROTHENDIECK's memoir on the subject [5]; compare also [15] and the literature given there.

2.5. Let H and K be Hilbert spaces. The algebraic tensor product of H and K may be given a positive definite hermitean bilinear form which is characterized by

$$(h \otimes k \mid h' \otimes k') = (h|h')_H (k|k')_K .$$

This makes the algebraic tensor product into a pre-Hilbert space. Its completion is called <u>the Hilbert space tensor product</u> and denoted by $H \otimes K$ (with a slight abuse of notation). The category of Hilbert spaces and bounded operators together with this tensor product forms a multiplicative category.

Naturally one wants to connect various multiplicative categories by functors.

2.6. <u>A functor</u> (or <u>cofunctor</u>) $S:(\mathcal{A},\otimes_{\mathcal{A}}) \longrightarrow (\mathcal{B},\otimes_{\mathcal{B}})$ <u>of multiplicative categories</u> is a functor (or a cofunctor) $S:\mathcal{A} \longrightarrow \mathcal{B}$ such that

$$\begin{array}{ccc} \mathcal{A} \times \mathcal{A} & \xrightarrow{\ \otimes_{\mathcal{A}}\ } & \mathcal{A} \\ {\scriptstyle S \times S}\downarrow & & \downarrow{\scriptstyle S} \\ \mathcal{B} \times \mathcal{B} & \xrightarrow[\ \otimes_{\mathcal{B}}\ ]{} & \mathcal{B} \end{array}$$

commutes. Instead of "functor of multiplicative categories" we will also speak of a "<u>multiplicative functor</u>".

Again let us look at a small selection out of the abundance of examples.

2.7. If $\mathfrak{A}$ and $\mathfrak{B}$ are categories with products (resp. coproducts) and if $S:\mathfrak{A} \longrightarrow \mathfrak{B}$ preserves finite products (resp., coproducts), then $S:(\mathfrak{A},\times) \longrightarrow (\mathfrak{B},\times)$ is a multiplicative functor.

This applies in particular to a multitude of product preserving forgetful functors.

There are a few interesting and less trivial examples.

2.8. Let $\mathfrak{G}^R$ be the category of R-modules, let $\wedge$ and P be the functors $\mathfrak{G}^R \longrightarrow \text{Grad}\,\mathfrak{G}^R$ which associate with an R-module the underlying graded module of its exterior algebra (such that all module elements have degree 1), respectively the underlying graded module of its symmetric algebra (such that all module elements have degree two). (The stipulation on the generators is of no relevance other than the fact that in this fashion the symmetric and the exterior algebra appear as special cases of one and the same type of graded algebras satisfying the commutation rule (which is best known in the case of the exterior algebra) $a^p b^q = (-1)^{pq} b^q a^p$ for homogeneous elements of degree p and q.) Then $\wedge, P:(\mathfrak{G}^R,\oplus) \longrightarrow (\text{Grad}\,\mathfrak{G}^R,\otimes^*_R)$ are functors of multiplicative categories. This is an expression of the fact that $\wedge(A\oplus B)$ and $\wedge A \otimes \wedge B$ are naturally isomorphic and the same holds with P in place of $\wedge$. It would be better for obvious reasons to replace the name of multiplicative functor by "exponential functor" as was done in [10].

2.9. Let R be a principal ideal domain, Comp the category of compact spaces and Comp_R the full subcategory of "R-torsion free spaces", i.e. spaces X for which the Čech-Alexander-Sheaf-cohomology modules $H^i(X;R)$ are flat. (Note that $\mathrm{Comp}_R = \mathrm{Comp}$ if R is a field.) Then Comp_R has finite products and the functor $H:(\mathrm{Comp}_R,\times) \longrightarrow (\mathrm{Grad}\,\mathcal{G}^R,\otimes_R^*)$ is multiplicative. This is a direct consequence of the definitions and the Künneth theorem.

The following two examples are illustrative with respect to some of our later discussions.

2.10. Let $\mathfrak{M}$ be the category of semigroups. Let A be a commutative ring with identity and $\mathfrak{A}$ be the category of A-algebras with identity. Denote with A[G] for a semigroup G the semigroup algebra (i.e. the R-algebra of all finitely valued functions $\varphi:G \longrightarrow A$ under pointwise addition and scalar multiplication, and under convolution $\varphi * \psi$ as multiplication, where $(\varphi * \psi)(x) = \Sigma\{\varphi(y)\psi(z):x = yz\}$). Then there is a natural isomorphism $A[G\times H] \longrightarrow A[G] \otimes_A A[H]$ under which the element (x,y) is sent onto the element $x \otimes_A y$ (where x is identified with the function $G \longrightarrow A$ taking the value 0 except in x where it takes the value 1). We observe therefore that $A[-]:(\mathfrak{M},\times) \longrightarrow (\mathfrak{A},\otimes_A)$ is a multiplicative functor.

A dual counterpart to the previous example is the following:

2.11. Let again $\mathfrak{M}$ be the category of semigroups and let $\mathfrak{M}_f$ be the full subcategory of finite semigroups. One observes that the functions $G \longrightarrow \mathbb{C}$ whose translates under left and right multiplication of the argument span a finite dimensional vector space form a subring $R(G)$ of $\mathbb{C}^G$. One can show that the natural isomorphism $\mathbb{C}^{G\times H} \longrightarrow \mathbb{C}^G \otimes_{\mathbb{C}} \mathbb{C}^H$ $\mathbb{C}^G \otimes_{\mathbb{C}} \mathbb{C}^H \longrightarrow \mathbb{C}^{G\times H}$ which associates with $\varphi \otimes \psi$ the function $(x,y) \longrightarrow \varphi(x)\psi(y)$ induces an isomorphism $R(G) \otimes_{\mathbb{C}} R(H) \longrightarrow R(G\times H)$. We thus note that $R(-)\colon(\mathfrak{M}_f,\times) \longrightarrow (\mathfrak{A},\otimes_{\mathbb{C}})$ is a multiplicative cofunctor.

Let us complement our list of examples by one which is of great significance in the theory of Lie algebras.

2.12. Let $\mathfrak{L}$ be the category of Lie algebras over a fixed field K. Let $\mathfrak{A}$ be the category of filtered associative K-algebras with identity and let $U\colon \mathfrak{L} \longrightarrow \mathfrak{A}$ be the functor which associates with any Lie algebra L its universal enveloping algebra. One shows that there is a natural isomorphism $U(L\times L') \longrightarrow U(L) \otimes_K U(L')$ which sends (x,y) onto $x \otimes 1 + 1 \otimes y$. Thus U is a multiplicative functor.

3. Algebras over a category with multiplication

We now utilize the concept of a multiplicative category to define the concept of a ring over a category.

3.1. DEFINITION. Let $(\mathcal{A},\otimes)$ be a multiplicative category. An algebra over $(\mathcal{A},\otimes)$ is an element m of $\mathcal{A}(A\otimes A, A)$ making the following diagram commute:

$$\begin{array}{ccc} A\otimes A\otimes A & \xrightarrow{m\otimes A} & A\otimes A \\ {\scriptstyle A\otimes m}\downarrow & & \downarrow{\scriptstyle m} \\ A\otimes A & \xrightarrow[m]{} & A \end{array}$$

If E is an identity for the category $(\mathcal{A},\otimes)$ we say that the algebra m has an identity if there is a morphism $\epsilon : E \longrightarrow A$ and a commutative diagram

$$\begin{array}{ccccc} E\otimes A & \xrightarrow{\epsilon\otimes A} & A\otimes A & \xleftarrow{A\otimes\epsilon} & A\otimes E \\ & {\scriptstyle\cong}\searrow & \downarrow{\scriptstyle m} & \swarrow{\scriptstyle\cong} & \\ & & A & & \end{array}$$

with the canonical isomorphisms defining the identity E. An algebra m over $(\mathcal{A},\otimes)$ is called commutative, if the diagram

$$\begin{array}{ccc} A\otimes A & \xrightarrow{t} & A\otimes A \\ & {\scriptstyle m}\searrow \quad \swarrow{\scriptstyle m} & \\ & A & \end{array}$$

is commutative where t is the automorphism of $A \otimes A$ obtained from 2.1.b via

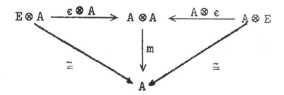

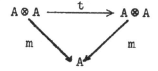

$$\begin{array}{ccc} A \times A & = & T(A \times A) \\ \otimes \downarrow & & \downarrow \otimes \\ A \otimes A & \xrightarrow[\cong]{t} & A \otimes A \end{array}$$

Let us look now at a few examples.

3.2. a) Let $\mathcal{S}$ be the category of sets and take the cartesian product as multiplication on $\mathcal{S}$. Then an algebra m over $(\mathcal{S}, \times)$ is obviously just a semigroup $m: A \times A \longrightarrow A$. It has an identity in the sense of definition 3.1 iff it has an identity in the conventional sense. It is commutative in the sense of definition 3.1 iff it is commutative in the conventional sense.

b) This example immediately carries over to the category $\mathcal{H}$ of topological spaces in place of sets. An algebra over $(\mathcal{H}, \times)$ then is a topological semigroup, and the remaining statements prevail.

3.3. Let $\mathcal{G}$ be the category of abelian groups and $\otimes$ the ordinary tensor product. An algebra m over $(\mathcal{G}, \otimes)$ then is an algebra (over $\mathbb{Z}$) in the conventional sense where we define the ring multiplication by $xy = m(x \otimes y)$. We observe that the use of the concept of an identity and of commutativity as introduced in definition 3.1 agrees with the conventional usage.

3.3.a. Let $\mathcal{G}^R$ be the category of R-modules over a commutative ring R with identity. Then an algebra m over

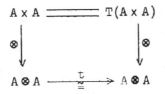

(Grad $\mathcal{G}^R, \otimes_R^*$) is exactly what ordinarily is called a graded R-algebra. Commutativity in this case is usually defined as follows and yields what traditionally was called anticommutativity: Let

$$T^* : \text{Grad } \mathcal{G}^R \times \text{Grad } \mathcal{G}^R \longrightarrow \text{Grad } \mathcal{G}^R \times \text{Grad } \mathcal{G}^R$$

be the functor which is defined by $T(f^p, g^q) = (-1)^{pq}(g^q, f^p)$ for $p, q \in \mathbb{Z}$. Let t^* be the automorphism of $A \otimes A$ obtained via

$$\begin{array}{ccc} A \times A & = & T(A \times A) \\ \otimes \downarrow & & \downarrow \otimes \\ A \otimes_R^* A & \xrightarrow{t^*} & A \otimes_R^* A \end{array}$$

Then an algebra m is called commutative if

$$\begin{array}{ccc} A \otimes_R^* A & \xrightarrow{t^*} & A \otimes_R^* A \\ & \searrow \quad \swarrow & \\ & A & \end{array}$$

commutes.

The exterior algebra $\wedge M$ and the symmetric algebra PM over an R-module M as described in example 2.8 are both commutative in this sense.

We observe that the codiagonal map $A \oplus A \longrightarrow A$ induces the algebra multiplication $\wedge A \otimes \wedge A \longrightarrow \wedge A$. A similar statement holds for the symmetric algebra.

Another example is the following:

3.4. Let $\mathcal{B}$ be the category of Banach spaces and $\hat{\otimes}$ the projective tensor product considered in 2.4. An algebra m over $(\mathcal{B},\hat{\otimes})$ is then a topological algebra with a multiplication defined by $xy = m(x\hat{\otimes}y)$ satisfying $\|xy\| \leq \|m\|\ \|x\hat{\otimes}y\| \leq \|m\|\ \|x\|\ \|y\|$; upon introducing the equivalent norm $|\ \ | = \|m\|\ \|\ \ \|$ we obtain a Banach algebra $(A,|\ \ |)$.

There is no problem in actually creating a category of algebras over a given multiplicative category $(\mathfrak{A},\otimes)$. In fact we define a morphism $\varphi:m \longrightarrow n$ of algebras over $(\mathfrak{A},\otimes)$ to be an element $\varphi \in \mathfrak{A}(A,B)$ such that the diagram

$$\begin{array}{ccc} A\otimes A & \xrightarrow{\varphi\otimes\varphi} & B\otimes B \\ {\scriptstyle m}\downarrow & & \downarrow{\scriptstyle n} \\ A & \xrightarrow[\varphi]{} & B \end{array}$$

commutes.

3.5. <u>DEFINITION</u>. The category which we have introduced is called <u>the category of algebras over</u> $(\mathfrak{A},\otimes)$. It is denoted with $\mathrm{alg}[(\mathfrak{A},\otimes)]$ or by $\mathrm{alg}[\mathfrak{A}]$ if no confusion is possible.

In concluding our list of examples let us look at algebras over the category of rings in the conventional sense.

3.6. The tensor product of the additive groups of two rings may be given the structure of a ring in a standard fashion so that $(a\otimes b)(c\otimes d) = ac\otimes bd$. Thus the category $\mathcal{R}$ of rings together with the ordinary tensor product is a multiplicative category. Suppose now that $m:A\otimes A \longrightarrow A$ is an algebra over the category $(\mathcal{R},\otimes)$. Then m is in fact a morphism of rings with identity in the conventional sense, i.e. we have

$$m(ac \otimes bd) = m(a \otimes b)\, m(c \otimes d).$$

In particular, if A has an identity and we define functions $\varphi,\psi:A \longrightarrow A$ by $\varphi(a) = m(a\otimes 1)$ and $\psi(b) = m(1\otimes b)$, then φ and ψ are ring endomorphisms preserving the identity such that

$$m(x \otimes y) = \varphi(x)\psi(y) = \psi(y)\varphi(x),\ x,y \in A.$$

From the associativity of m is follows further that

$$\varphi^2 = \varphi,\ \psi^2 = \psi,\ \text{and}\ \ \varphi\psi = \psi\varphi$$

i.e. that φ and ψ are two commuting projections (idempotent endomorphisms). Conversely, if φ and ψ are endomorphisms of the ring A preserving the identity and satisfying all the above properties, then there is a unique ring morphism preserving the identity $m:A\otimes A \longrightarrow A$ with $m(x\otimes y) = \varphi(x)\psi(y)$.

The ring $\mathbb{Z}$ is the identity of the multiplicative category $(\mathcal{R},\otimes)$; if the algebra $m:A\otimes A \longrightarrow A$ over $(\mathcal{R},\otimes)$ has an identity, then there is a morphism $\varepsilon: \mathbb{Z} \longrightarrow A$ in $\mathcal{R}$ such that $m(1\otimes a) = m(\varepsilon(1)\otimes a) = \varepsilon(1)a = 1a = a$ and similarly $m(a\otimes 1) = a$. This means that $\varphi = \psi =$ identity of A, and that m is in fact the given multiplication, and A is a commutative ring. Thus an algebra with identity over $(\mathfrak{R},\otimes)$ is in fact nothing but a commutative ring.

The reader is invited to carry the program of the preceding example 3.6 through for the category of Banach algebras with identity and continuous identity preserving algebra morphisms in place of the category $\mathfrak{R}$ and with the projective tensor product $\hat{\otimes}$ in place of $\otimes$. He will then arrive at the following:

3.7. Let $\mathfrak{A}$ be the category of Banach algebras with identity and $\hat{\otimes}$ the projective tensor product (where the projective tensor product of the underlying Banach spaces of two algebras has the obvious Banach algebra structure). Then $(\mathfrak{A},\hat{\otimes})$ is a multiplicative category, and the category of algebras with identity over $(\mathfrak{A},\hat{\otimes})$ may be naturally identified with the category of commutative Banach algebras with identity.

The tensor product of rings in the conventional sense may without difficulty be generalized to the categorical concept of a ring.

3.8. Let $(\mathfrak{A},\otimes)$ be a multiplicative category and $m:A\otimes A \longrightarrow A$ an algebra over $(\mathfrak{A},\otimes)$. Let $\varphi_{B,C,D,E}:B\otimes C\otimes D\otimes E \longrightarrow B\otimes D\otimes C\otimes E$ be the natural isomorphism derived from the commutativity of the product 2.1.b. Then the element $m \otimes' m \in \mathfrak{A}(A\otimes A\otimes A\otimes A, A\otimes A)$ defined by the diagram

$$\begin{array}{ccc} A \otimes A \otimes A \otimes A & \xrightarrow[\varphi_{A,A,A,A}]{} & A \otimes A \otimes A \otimes A \\ & {\scriptstyle m \otimes' m} \searrow \quad \swarrow {\scriptstyle m \otimes m} & \\ & A \otimes A & \end{array}$$

is an algebra over the category $(\mathcal{O}\!\mathcal{l},\otimes)$.

The category of algebras over $(\mathcal{O}\!\mathcal{l},\otimes)$ thus becomes a multiplicative category itself relative to the multiplication $\otimes'$. One can determine now in purely categorical terms, under what circumstances the repetition of this construction becomes stationary; the ideas are already discussed in the concrete case of 3.6.

Let us observe the straightforward fact that multiplicative functors preserve algebras:

3.9. Let $(\mathcal{O}\!\mathcal{l},\otimes_{\mathcal{O}\!\mathcal{l}})$ and $(\mathcal{B},\otimes_{\mathcal{B}})$ be multiplicative categories. Let $T\colon \mathcal{O}\!\mathcal{l} \longrightarrow \mathcal{B}$ be a multiplicative functor. If $m\colon A \otimes_{\mathcal{O}\!\mathcal{l}} A \longrightarrow A$ is an algebra over $(\mathcal{O}\!\mathcal{l},\otimes_{\mathcal{O}\!\mathcal{l}})$, then $Tm\colon TA \otimes \;\; TA \longrightarrow TA$ is an algebra over $(\mathcal{B},\otimes_{\mathcal{B}})$. More accurately, T induces a functor from the category of algebras over $(\mathcal{O}\!\mathcal{l},\otimes_{\mathcal{O}\!\mathcal{l}})$ into the category of algebras over $(\mathcal{B},\otimes_{\mathcal{B}})$.

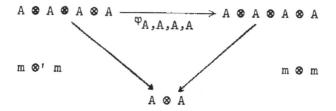

There is, naturally a certain question as to the adequacy of the term "algebra" in the general context which we have described. If, as we did, we consider example 3.3 as the motivating one, then the nomenclature is unquestionably adequate. If, however, example 3.2 is considered to be the typical one, then the term "semigroup" would certainly be better suited.

4. Cogebras over a category with multiplication

There is, apart from the abundance of traditional examples of algebras which we indicated, no essential reason why one ought to single out morphisms $A \otimes A \longrightarrow A$ and not morphisms $A \longrightarrow A \otimes A$. Only in comparatively recent years has the significance of the latter concept gained recognition, mostly through the conditioning towards more categorical thinking. The formal process of defining the co-concepts introduced in the previous section is the usual.

We make the following trivial observation:

4.1. If $(\mathfrak{A},\otimes)$ is a multiplicative category, then $(\mathfrak{A}^{\circ},\otimes)$ is a multiplicative category, where $\mathfrak{A}^{\circ}$ is the opposite category of $\mathfrak{A}$ (1.2). Any identity of $(\mathfrak{A},\otimes)$ is an identity of $(\mathfrak{A}^{\circ},\otimes)$.

4.2. DEFINITION. Let $(\mathfrak{A},\otimes)$ be a multiplicative category. Then a cogebra over $(\mathfrak{A},\otimes)$ is an algebra over $(\mathfrak{A}^{\circ},\otimes)$. Notably, a cogebra is an element $m \in \mathfrak{A}(A,A\otimes A)$; the commutativity of the dual of the diagram in 3.1 is called the coassociativity of m.

Suppose that $(\mathfrak{A},\otimes)$ has an identity E. A cogebra $m:A \longrightarrow A \otimes A$ has a coidentity $\varepsilon:A \longrightarrow E$ if ε is an identity of m in $(\mathfrak{A}^\circ,\otimes)$. A cogebra is called cocommutative, if it is a commutative algebra in $\mathfrak{A}^\circ$. The category of cogebras over $(\mathfrak{A},\otimes)$ simply is defined to be the category of algebras over $(\mathfrak{A}^\circ,\otimes)$, and it is denoted by co$[\mathfrak{A}]$.

The following observation is in complete analogy to the dual one made about multiplicative functors and their action on categories of rings:

4.3. Let $(\mathfrak{A},\otimes_{\mathfrak{A}})$ and $(\mathfrak{B},\otimes_{\mathfrak{B}})$ be multiplicative categories. Let T be a multiplicative functor. If $m:A \longrightarrow A \otimes_{\mathfrak{A}} A$ is a cogebra over $(\mathfrak{A},\otimes_{\mathfrak{A}})$, then $Tm:TA \longrightarrow TA \otimes_{\mathfrak{B}} TA$ is a cogebra over $(\mathfrak{B},\otimes_{\mathfrak{B}})$. More accurately, T induces a functor from the category of cogebras over $(\mathfrak{A},\otimes_{\mathfrak{A}})$ into the category of over $(\mathfrak{B},\otimes_{\mathfrak{B}})$.

This is just a repetition of 3.9. It may happen, however, that only one of the categories involved in 3.9 is replaced by its opposite category. This then results in the following analogue of 3.9.

4.4. A multiplicative cofunctor from a multiplicative category $(\mathfrak{A},\otimes_{\mathfrak{A}})$ into another multiplicative category $(\mathfrak{B},\otimes_{\mathfrak{B}})$ transforms algebras into cogebras and vice versa. More accurately, a multiplicative functor induces a cofunctor from the category of algebras [resp., cogebras] over $(\mathfrak{A},\otimes_{\mathfrak{A}})$ into the category of cogebras [resp., algebras] over $(\mathfrak{B},\otimes_{\mathfrak{B}})$.

Let us now look at a selection of concrete examples of cogebras. A trivial one first:

4.5. Let $(\mathcal{S},\times)$ be the category of sets with the Cartesian product as multiplication. For any set X, the diagonal map $d_X : X \longrightarrow X \times X$ (defined by $d_X(x) = (x,x)$) is a cocommutative cogebra. The multiplicative category $(\mathcal{S},\times)$ has any singleton set E as identity, and the constant function $\epsilon : X \longrightarrow E$ is a coidentity. In fact, there is only one cogebra structure on X with these properties: Set $c(x) = (\varphi(x),\psi(x))$. One deduces readily from the definitions that φ and ψ are functions satisfying $\varphi^2 = \varphi$, $\varphi\psi = \psi\varphi$, $\psi^2 = \psi$. Clearly c is cocommutative iff $\varphi = \psi$. It has a coidentity iff $\varphi = \psi$ = identity of X.

This example may in fact be generalized almost indefinitly without ridding it of its essential triviality; instead of the category of sets we may take any category with finite products and still obtain a cocommutative cogebra over the multiplicative category. However, the situation attains a somewhat more interesting flavor as soon as we observe, that some less trivial examples are obtained by applying multiplicative functors to the situations above. In fact I do not know of any naturally occuring cogebras which do not in one way or another arise in this fashion.

4.6. Let $\mathcal{G}^R$ be the category of R-modules and $\Lambda, P : \mathcal{G}^R \longrightarrow \text{Grad}\ \mathcal{G}^R$ the multiplicative functors described in 2.8. For an R-module A consider the cogebra

$d_A : A \longrightarrow A \oplus A$ over $(\mathcal{G}^R, \oplus)$. Then

$$\wedge A \xrightarrow{\wedge d_A} \wedge(A \oplus A) \xrightarrow{\cong} \wedge A \otimes^*_R \wedge A$$

and

$$PA \longrightarrow P(A \oplus A) \xrightarrow{\cong} PA \otimes^*_R A$$

are cogebras over $(\mathrm{Grad}\,\mathcal{G}^R, \otimes^*_R)$.

4.7. Let R be a principal ideal domain, Comp the category of compact spaces and Comp_R the full subcategory of R-torsion free spaces (see 2.9). An algebra $m : S \times S \longrightarrow S$ over $(\mathrm{Comp}_R, \times)$ is a compact topological semigroup with R-torsion free underlying space (see 3.2.6). Let $H : \mathrm{Comp}_R \longrightarrow \mathrm{Grad}\,\mathcal{G}^R$ be the Čech cohomology functor as in 2.9. Then

$$HS \xrightarrow{Hm} H(S \times S) \xrightarrow{\cong} HS \otimes^*_R HS$$

is a cogebra over $(\mathrm{Grad}\,\mathcal{G}^R, \otimes)$.

Note that all this applies to any compact semigroup if R is a field. It has been shown that any compact abelian semigroup has a $\mathbb{Z}$-torsion free underlying space. In fact, if S is a compact connected abelian group, then $HS \cong \wedge\hat{S}$ both as a graded algebra and cogebra where S denotes the character group of S [10].

4.8. Let A be a commutative ring with identity, G a semigroup. Let $d_G : G \longrightarrow G \times G$ be the diagonal cogebra over the category of semigroups with $\times$. Then

$$A[G] \xrightarrow{A[d_G]} A[G \times G] \xrightarrow{\cong} A[G] \otimes A[G]$$

is a cogebra over the category of A-algebras with ⊗.

4.9. If L is a Lie algebra over a field K and $d_L : L \longrightarrow L \times L$ the diagonal cogebra, then

$$U(L) \xrightarrow{U(d_L)} U(L \times L) \xrightarrow{\cong} U(L) \otimes U(L)$$

is a cogebra over the category of filtered K-algebras with ⊗ (see 2.10).

5. Bigebras

Some of the most interesting and significant cogebra are in fact called bigebras (or, a little more traditionally, Hopf algebras or hyperalgebras). Why the distinction? The exact definition of a bigebra (which is suitable to cover most applications) is the following:

5.1. DEFINITION. Let $(\mathfrak{A},\otimes)$ be a multiplicative category. A bigebra over $(\mathfrak{A},\otimes)$ (based on the object A) is a diagram

$$A \xrightarrow{\ c\ } A \otimes A \xrightarrow{\ m\ } A$$

consisting of a cogebra c and an algebra m over $(\mathfrak{A},\otimes)$ such that the following diagram commutes:

$$\begin{array}{ccccc}
A \otimes A \otimes A \otimes A & & \xrightarrow{\ \varphi_{A,A,A,A}\ } & & A \otimes A \otimes A \otimes A \\
\uparrow{\scriptstyle c \otimes c} & & & & \downarrow{\scriptstyle m \otimes m} \\
A \otimes A & \xrightarrow[\ m\ ]{} & A & \xrightarrow[\ c\ ]{} & A \otimes A
\end{array}$$

with the involution $\varphi_{A,A,A,A}$ as in 3.8. This is equivalent to saying that c is a morphism of algebras although in this version the symmetry of the concept is not so obvious. A morphism $\varphi:(c,m) \longrightarrow (c',m')$ of bigebras based on the objects A and A', respectively, is an element $\varphi \in \mathfrak{A}(A,A')$ such that

$$\begin{array}{ccccc}
A & \longrightarrow & A \otimes A & \longrightarrow & A \\
\downarrow{\scriptstyle \varphi} & & \downarrow{\scriptstyle \varphi \otimes \varphi} & & \downarrow{\scriptstyle \varphi} \\
A' & \xrightarrow{\ c'\ } & A' \otimes A' & \xrightarrow{\ m'\ } & A'
\end{array}$$

commutes. Clearly one defines in this way a category, <u>the category of bigebras over</u> $(\mathcal{A}, \otimes)$. Note that any bigebra over $(\mathcal{A}, \otimes)$ is also a bigebra over $(\mathcal{A}^{o}, \otimes)$. The category of bigebras over $\mathcal{A}$ is denoted by bi$[\mathcal{A}]$.

The following is immediate:

5.2. If $(\mathcal{A}, \otimes_{\mathcal{A}})$ and $(\mathcal{B}, \otimes_{\mathcal{B}})$ are multiplicative categories and T: $\mathcal{A} \longrightarrow \mathcal{B}$ is a multiplicative functor [cofunctor], then

$$TA \underset{[\longleftarrow]}{\overset{Tc}{\longrightarrow}} TA \otimes TA \underset{[\longleftarrow]}{\overset{Tm}{\longrightarrow}} TA$$

is a bigebra (where TA ⊗ TA and T(A ⊗ A) are identified under the natural isomorphism defining the multiplicativity of T) provided that (c,m) is a bigebra. More accurately, the functor [cofunctor] T induces a functor [cofunctor] from the category of bigebras over $(\mathcal{A}, \otimes_{\mathcal{A}})$ into the category of bigebras over $(\mathcal{B}, \otimes_{\mathcal{B}})$.

The actual reason why so many cogebras which we have encountered as important examples are in fact bigebras is due essentially to the fact that they are cogebras over categories of commutative algebras with identity so that the part of the definition of the bigebra referring to the algebra structure is already implicit in the object A as was illustrated in example 3.6. In particular, any cogebra over the category of commutative rings is in fact a bigebra where the algebra structure is the one defined by the ring multiplication, and if the cogebra has an identity (i.e. the algebra on which the cogebra is defined has an identity and the comultiplication preserves identities) then the algebra structure m is unique (3.6).

5.3 DEFINITION. Let $A \xrightarrow{c} A \otimes A \xrightarrow{m} A$ be a bigebra over $(\mathfrak{A}, \otimes)$. It has an identity, if m has an an identity and c preserves identities. It is said to be commutative, if m is commutative, and cocommutative, if c is cocommutative. The bigebra (c,m) is said to be idempotent, if mc = the identity of A. It is said to have a coidentity if it has an identity in $\mathfrak{A}^o$.

Let us consider a few examples; in principle they all arise in a fashion similar to the one in which our examples of cogebras originated--which is no surprise after the preceding remark.

5.4. Let $\mathcal{Y}$ be the category of sets. Let $m: S \times S \longrightarrow S$ be a ring over $(\mathcal{Y}, \times)$, i.e. a semigroup. Then

$$S \xrightarrow{d_S} S \times S \xrightarrow{m} S$$

is a bigebra over $(\mathcal{Y}, \times)$. It is, in fact cocommutative and has a coidentity (namely the constant morphism $S \longrightarrow \{1\}$). Conversely, let us assume that

$$S \xrightarrow{c} S \times S \xrightarrow{m} S$$

is any bigebra. Set $c(s) = (\varphi(s), \psi(s))$. One deduces readily from the definition that φ and ψ a semigroup endomorphisms of S satisfying $\varphi^2 = \varphi$, $\varphi\psi = \psi\varphi$, $\psi^2 = \psi$. Clearly c is cocommutative if and only if $\varphi = \psi$. It has a coidentity iff $\varphi = \psi = S$. Thus given a semigroup m, the only bigebra structure based on S which is cocommutative and has a coidentity is the example given initially. It is in that loose sense that we may identify the category of semigroups with the category of cocommutative bigebras with coidentity over $(\mathcal{Y}, x)$. Note that (c,m) is idempotent iff S is an idempotent semigroup.

The preceding example immediately generalizes to topological semigroups, compact topological semigroups, etc.

5.5. For R-modules the bigebra

$$A \xrightarrow{d_A} A \oplus A \xrightarrow{m} A$$

with module addition m immediately gives rise to the graded commutative bigebras

$$\Lambda A \xrightarrow{\Lambda d_A} \Lambda A \otimes \Lambda A \xrightarrow{\Lambda m} \Lambda A$$

$$PA \xrightarrow{Pd_A} PA \otimes PA \xrightarrow{Pm} PA$$

In particular, every polynomial ring is a bigebra since every polynomial ring is a PA with a free module A.

5.6. For a principal ideal domain R and any R-torsion free compact topological semigroup S, the Čech-cohomology functor immediately gives rise to the graded commutative bigebra

$$HS \xrightarrow{Hm} HS \otimes HS \xrightarrow{U=Hd_S} HS$$

This example is the historical origin of the concept of a bigebra.

5.7. If L is a Lie algebra over a field K, then U(L) is a filtered K-algebra and U is a functor; thus $U(L) \xrightarrow{U(d_L)} UL \otimes UL$ is in fact an algebra morphism since $U(L \times L) \cong UL \otimes UL$ is an algebra morphism. Thus, if we write $m: UL \otimes UL \longrightarrow UL$ for the algebra multiplication, then

$$UL \xrightarrow{Ud_L} UL \otimes UL \xrightarrow{\quad m \quad} UL$$

is a bigebra. One should remark that in this case this is not derived from a bigebra $L \longrightarrow L \times L \longrightarrow L$ because there is no morphism $L \times L \longrightarrow L$ in the category of Lie algebras.

5.8. A similar situation prevails in the case of semigroup algebras. Let A be a commutative ring. The functor A[ ] maps the category of semigroup into the category of A-algebras. Thus the morphism $d_G: G \longrightarrow G \times G$ produces an algebra morphism $A[G] \longrightarrow A[G] \otimes A[G]$. Hence, if we denote the algebra multiplication $A[G] \times A[G] \longrightarrow A[G]$ (the convolution) with m, then we obtain indeed a bigebra

$$A[G] \xrightarrow{A[d_A]} A[G] \otimes A[G] \xrightarrow{\quad m \quad} A[G].$$

The map m is of the form A[n] with a semigroup morphism $n: G \times G \longrightarrow G$ if G is commutative, in which case we might write $n(g,h) = gh$.

PART II

6. C*-algebras

The mathematics of the previous section as such was superficial by comparison, although the special examples occasionally lead into deeper mathematical theories. The current section is devoted to the category of C*-algebras under the general aspects outlined in Section 2. Naturally we will lean on the theory of C*-algebras and use the information available in the literature most of which is excellently accessible through the exposition of this theory by DIXMIER [4].

We first recall the definition of a C*-algebra. In essence, a C*-algebra is a complex Banach algebra with an additional element of structure, an involution which is in a very strong way coupled with the norm. Specifically:

6.1. DEFINITION. Let A be a complex algebra. An *involution* of A is an automorphism $x \longrightarrow x^*$ of the additive group satisfying $x^{**} = x$ such that $(cx)^* = \bar{c}x^*$ for $x \in A$, $c \in \mathbb{C}$ where $\bar{c}$ is the complex conjugate of c, and that $(xy)^* = y^*x^*$ for all $x,y \in A$. The pair (A,*) is called *an involutive algebra*.

An element x of an involutive algebra which is fixed under the involution is called *selfadjoint* or *hermitean*. If A_h denotes the subvector space of hermitean elements, then $A = A_h \oplus iA_h$. In fact for any element $x \in A$ we have a unique decomposition in the form $x = x' + ix''$ with $x', x'' \in A_h$; specifically, let $x' = \frac{1}{2}(x + x^*)$ and $x'' = \frac{1}{2i}(x - x^*)$. The involution then simply takes the form $(x' + ix'')^* = x' - ix''$, i.e. A_h resp. iA_h are the eigenspaces for the eigenvalues 1 resp. -1 of $*$.

A morphism of involutive algebras is an algebra morphism f satisfying $f(x^*) = f(x)^*$ for all $x \in \operatorname{dom} f$. The concept of <u>the category of involutive algebras</u> is now clear. The category of involutive algebras is denoted with Inv.

6.2. <u>DEFINITION</u>. <u>An involutive Banach algebra</u> is a complex Banach algebra A which at the same time is an involutive algebra in which the involution is an isometry. <u>A C*-algebra</u> is an involutive Banach algebra such that the following identity holds for all $x \in A$:

$$\|x^*x\| = \|x\|^2 .$$

The characteristic examples are the following:

6.3. i) Let X be a compact Hausdorff space and $A = C(X)$, the algebra of all complex value functions with $f^*(x) = \overline{f(x)}$. Then A is a commutative C*-algebra with identity.

ii) Let X be a locally compact but not compact Hausdorff space and $A = C_o(X)$ the algebra of all complex valued functions on X vanishing at infinity with complex conjugation as involution. Then A is a commutative C*-algebra without identity.

iii) Let H be a complex Hilbert space and $A \subset B(H)$ a uniformly closed selfadjoint subalgebra of the algebra B(H) of all bounded operators on H with x* being the adjoint of the operator x. Then A is a C*-algebra.

It belongs to the fundamentals of the theory, that the first example exhausts all commutative C*-algebras with identity, the second all commutative C*-algebras without identity, and the third all C*-algebras. Later on we will mostly concentrate on the first category.

It is again a piece of standard information about C*-algebras that any algebra morphism between C*-algebras which preserves the involution is automatically continuous, in fact a contraction. In the same line it is true that every closed (two sided) ideal of a C*-algebra is automatically selfadjoint, i.e. stable under the involution.

6.4. <u>DEFINITION</u>. <u>The category</u> C* <u>of</u> C*-<u>algebras</u> has C*-algebras as objects and morphisms of C*-algebras as morphisms.

It is perhaps instructive to describe the nature of the category of commutative C*-algebras, in particular, since it is this subcategory which we will be using predominantly in the later parts.

<u>A character</u> of a C*-algebra A is a non-zero C*-algebra morphism $f:A \longrightarrow \mathbb{C}$. The set of all characters may at the same time be considered as the set of classes of irreducible

representations of A on a Hilbert space (all of which are one dimensional); this set is called the spectrum of A, and we denote it by Spec A. Since Spec A is a subset of the dual of the Banach space underlying A, it may be given the topology of pointwise convergence, the so-called weak star topology. This topology is always locally compact Hausdorff, and is in fact compact if and only if A has an identity. There is a natural bijective correspondence between the characters and the maximal modular ideals (the latter being exactly the kernels of the former); recall that all maximal modular ideals are primitive. As a space of primitive, hence prime ideals we have another canonical topology, namely the hull-kernel topology. Fortunately in the present situation there is no need for the distinction, since both topologies agree.

If $a \in A$ then the function $\hat{a}$:Spec A $\longrightarrow \mathbb{C}$ defined by $\hat{a}(f) = f(a)$ is in fact continuous and vanishes at infinity if A has no identity. Let C(X) for any space X be the C*-algebra of all bounded continuous functions $f: X \longrightarrow \mathbb{C}$. Let $C_o(X)$ for a locally compact space either be C(X) if X is compact or the subalgebra of all continuous functions vanishing at ∞ otherwise. The function $a \longrightarrow \hat{a}: A \longrightarrow$ C_o(Spec A) is an isomorphism in C*, the so-called Gelfand isomorphism.

If $\varphi: A \longrightarrow B$ is a morphism in C* and $f \in$ Spec B, then $f\varphi \in$ Spec $A \cup \{0\}$. Let us suppose that $\varphi(A)$ is not contained in any maximal modular ideal of B. Then $f \longmapsto f\varphi$: Spec B $\longrightarrow$ Spec A is a well-defined function Spec φ. It is clearly continuous but has in fact an additional property: Recall that a continuous function between locally compact spaces is called proper, if it is closed and the inverse image of each point of the codomain is compact; we will show that Spec φ is proper. Suppose that $f \in$ Spec A; let $K = \bigcap \{\ker g : g \in \text{Spec } B \text{ and } g\varphi = f\}$; now $(\text{Spec } \varphi)^{-1}(f)$ is compact iff B/K has an identity. But let $e \in A$ be such that $f(e) = 1$ and $g\varphi = f$ with $g \in$ Spec B; then $g(b - b\varphi(e)) = g(b) - g(b)g\varphi(e) = g(b) - g(b)f(e) = 0$; so $\varphi(e) + K$ is an identity for A/K. It remains to see that Spec φ is closed. This is clear if φ is surjective ([4],p.61). The case that φ is injective remains, since φ has a surjective-injective factorisation. We may now assume that A is a subalgebra of B which is not annihilated by any character of B. Let $\tilde{B}$ be the C*-algebra obtained from B by adjoining an identity ([4],p.7) and likewise $\tilde{A} = A + \mathbb{C}.\tilde{1} \subset \tilde{B}$. Then Spec $\tilde{B}$ = Spec B $\cup$ {e} where e is the character given by $e(b + c.\tilde{1}) = c$, and Spec $\tilde{A}$ = Spec A $\cup$ $\{e|A\}$. Clearly no character of $\tilde{B}$ is annihilated on $\tilde{A}$. If $X \subset$ Spec B is closed then $X \cup \{e\}$ is compact; if $Y = \{f|A: f \in X\}$ then $Y \cup \{e|A\}$ is compact as image of $X \cup \{e\}$; but then $Y = (Y \cup \{e|A\}) \cap$ Spec A is closed. This finishes the argument.

Conversely, let $\psi: X \longrightarrow Y$ be a proper map of locally compact spaces; if we define $C_o(\psi): C_o(Y) \longrightarrow C_o(X)$ by $C_o(\psi)(f) = f\psi$, then $C_o(\psi)$ is a morphism of C*-algebras such that im $C_o(\psi)$ is not contained in any maximal ideal of

$C_o(X)$. Firstly observe that for any $g \in C_o(Y)$ the function $g\psi$ is not only continuous, but vanishes at infinity, since ψ is proper; and secondly for each $x \in X$, there is a $g \in C_o(Y)$ such that $g\psi(x) \neq 0$, whence im $C_o(\psi)$ cannot be contained in any maximal modular ideal.

This leads us to the following definition:

6.5. DEFINITION. A morphism $\varphi: A \longrightarrow B$ of commutative C*-algebras is called proper if $\varphi(A)$ is not contained in any maximal modular ideal of B (i.e. is not annihilated by any character of B).

Note that any identity preserving morphism of commutative C*-algebras with identity is automatically proper. Let $\mathcal{L}$ be the category of locally compact spaces and proper maps, and C^*_{prop} the category of commutative C*-algebras with proper morphisms. Let C^*_a be the category of commutative C*-algebras with identity, and Comp the category of compact spaces.

We then have the following result, which gives us a prime example for a duality theory (see Section 1):

6.6. THEOREM (GELFAND and NAIMARK).

i) The functors $C_o : \mathcal{L} \longrightarrow C^*_{prop}$

and $\mathrm{Spec} : C^*_{prop} \longrightarrow \mathcal{L}$

form a duality. That is, the categories $\mathcal{L}$ and C^*_{prop} are dual.

ii) <u>The functors</u> $C : \text{Comp} \longrightarrow C^*_a$ <u>and</u> $\text{Spec} : C^*_a \longrightarrow \text{Comp}$ <u>form a duality. That is, the categories</u> Comp <u>and</u> C^*_a <u>are dual.</u>

iii) <u>The dualities given in</u> i) <u>and</u> ii) <u>are compatible.</u> (Sec. 1.3.)

7. <u>A multiplication for the category</u> C*

We are now going to introduce a tensor product for C*-algebras. We start with two technical lemmas.

7.1. <u>LEMMA</u>. Let A_i, $i = 1,2$ be complex vector spaces, V a real vector space, and $f:A_1 \times A_2 \longrightarrow V$ a real bilinear map such that $f(ca_1,a_2) = f(a_1,ca_2)$ for $a_i \in A_i$, $i = 1,2$, $c \in \mathbb{C}$. Then there is a unique real vector space morphism $f':A_1 \otimes_{\mathbb{C}} A_2 \longrightarrow V$ such that $f'(a_1 \otimes_{\mathbb{C}} a_2) = f(a_1,a_2)$.

Proof. If $f'':A_1 \otimes_{\mathbb{R}} A_2 \longrightarrow V$ is the unique morphism with $f''(a_1 \otimes_{\mathbb{R}} a_2) = f(a_1,a_2)$ then by hypotheses f'' vanishes on the subspace I spanned by all $ca_1 \otimes_{\mathbb{R}} a_2 - a_1 \otimes_{\mathbb{R}} ca_2$, $a_i \in A_i$, $c \in \mathbb{C}$. But $(A_1 \otimes_{\mathbb{R}} A_2)/I \cong A_1 \otimes_{\mathbb{C}} A_2$, whence the assertion

7.2. <u>LEMMA</u>. Let A_i, $i = 1,2$ be involutive vector spaces (i.e. complex vector spaces with an involution satisfying $(ca)^* = \bar{c}a^*$). Then $A_1 \otimes_{\mathbb{C}} A_2$ has a unique involution satisfying $(a_1 \otimes_{\mathbb{C}} a_2)^* = a_1^* \otimes_{\mathbb{C}} a_2^*$.

Proof. We apply Lemma 7.1 to $f: A_1 \times A_2 \longrightarrow A_1 \otimes_{\mathbb{C}} A_2$ with $f(a_1, a_2) = a_1{}^* \otimes_{\mathbb{C}} a_2{}^*$. There is then a unique real morphism $*: A_1 \otimes_{\mathbb{C}} A_2 \longrightarrow A_1 \otimes_{\mathbb{C}} A_2$ such that $(a_1 \otimes_{\mathbb{C}} a_2)^* = a_1{}^* \otimes_{\mathbb{C}} a_2{}^*$. It follows readily that $*$ on $A_1 \otimes_{\mathbb{C}} A_2$ is an involution and satisfies $(cx)^* = \bar{c}x^*$ for all $x \in A_1 \otimes_{\mathbb{C}} A_2$.

Convention. From here on we will write $\otimes$ in place of $\otimes_{\mathbb{C}}$ for the tensor product of complex vector spaces.

7.3. If A_i, $i = 1,2$ are involutive algebras, then $A_1 \otimes A_2$ is an involutive algebra in such a fashion, that $(a_1 \otimes a_2)^* = a_1{}^* \otimes a_2{}^*$. The pair $(\text{Inv}, \otimes)$ is a multiplicative category.

7.4. Let A_1 and A_2 be two involutive Banach algebras. The projective tensor product $A_1 \hat{\otimes} A_2$ is again complex Banach algebra containing the involutive algebra $A_1 \otimes A_2$. With arguments completely analogous to the ones in 7.1 through 7.3 it follows that $A \hat{\otimes} B$ has an involution of Banach algebras satisfying $(a \hat{\otimes} b)^* = a^* \hat{\otimes} b^*$. Thus the projective tensor product with the additional structure given by the involution is a multiplication on the category of involutive algebras.

We have to recall that there is a coreflection from the category of involutive Banach algebras into the category of C^*-algebras; indeed, given any involutive Banach algebra A there is a C^*-algebra $\tilde{A}$ and a natural morphism $\varphi_A : A \longrightarrow \tilde{A}$ of involutive Banach algebras such that any morphism $\psi : A \longrightarrow B$ of involutive Banach algebras into a C^*-algebra

B factors uniquely through φ_A. The algebra $\tilde{A}$ is called *the enveloping C*-algebra* of A. (For an explicit construction not using the adjoint functor theorem the reader is referred to DIXMIER [4].)

7.5. DEFINITION. Let A_1 and A_2 be two C*-algebras. We define *the C*-algebra tensor product* $A_1 \otimes^* A_2$ to be $(A_1 \hat{\otimes} A_2)^{\sim}$. In particular, there is a natural morphism of involutive algebras $\varphi_{A_1A_2}: A_1 \hat{\otimes} A_2 \longrightarrow A_1 \otimes^* A_2$. The associativity and commutativity of $\otimes^*$ is readily checked from the definitions, using the universal property of the enveloping C*-algebra.

Let us list this observation as

7.6. *The pair* $(C^*, \otimes^*)$ *is a multiplicative category*. From the construction of the projective tensor product we know that the algebraic tensor product of the two C*-algebras A_1 and A_2 is contained in $A_1 \hat{\otimes} A_2$; we also know the existence of a canonical map of $A_1 \hat{\otimes} A_2$ into $A_1 \otimes^* A_2$ with a dense image; question: is $A_1 \otimes A_2$ injected into $A_1 \otimes^* A_2$ so that in fact we may consider the algebraic tensor product as a dense involutive subalgebra of the C*-algebra $A_1 \otimes^* A_2$? The answer is

7.7. *The natural map* $i_{A_1A_2}: A_1 \otimes A_2 \longrightarrow A_1 \otimes^* A_2$ *is injective*.

Proof. Since A_1 and A_2 have faithful representations on Hilbert spaces we may assume that A_1 is a uniformly closed selfadjoint subalgebra of the algebra $B(H_i)$ of a Hilbert space H_i. We form the Hilbert space tensor product $H_1 \otimes H_2$. (See 2.5.) The algebraic tensor product $A_1 \otimes A_2$ may be identified with a selfadjoint subalgebra of $B(H_1 \otimes H_2)$ satisfying $(\varphi_1 \otimes \varphi_2)(h_1 \otimes h_2) = \varphi_1(h_1) \otimes \varphi_2(h_2)$, so that $(\varphi_1, \varphi_2) \longrightarrow \varphi_1 \otimes \varphi_2 : A_1 \times A_2 \longrightarrow B(H_1 \otimes H_2)$ is a continuous bilinear map. It therefore factors uniquely through a map $F: A_1 \hat{\otimes} A_2 \longrightarrow B(H_1 \otimes H_2)$ which in turn, since $B(H_1 \otimes H_2)$ is a C*-algebra, and F is a morphism of involutive Banach algebras, factors through $A_1 \otimes^* A_2$. We therefore have a sequence of maps

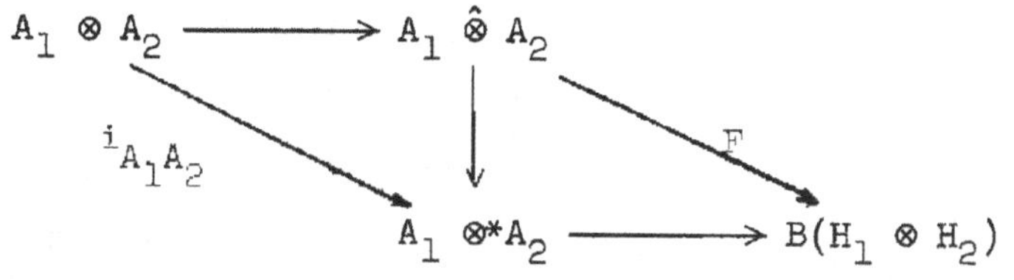

whose composition we observed to be an injection. This shows that $i_{A_1A_2}$ is injective.

We inspect the definition of $\otimes^*$ a little more carefully and compare its relation to previous definitions.

By the construction of the universal C*-enveloping algebra of an involutive Banach algebra ([4], p. 40) the tensor product $A_1 \otimes^* A_2$ is the completion of $A_1 \hat{\otimes} A_2$ relative to the norm given by

$\|x\| = \sup\{ \|\pi(x)\| : \pi$ is any $*$-representation of $A_1 \hat{\otimes} A_2$ on a Hilbert space$\}$

$= \sup\{ |\pi(x)| : \pi$ is any irreducible $*$-representation of $A_1 \hat{\otimes} A_2$ on a Hilbert space$\}$.

Now there is a bijective correspondence between the $*$-representations of $A_1 \hat{\otimes} A_2$ on one hand and the $*$-representations of $A_1 \otimes A_2$ satisfying $\|\pi(a_1 \otimes a_2)\| \leqq \|a_1\| \; \|a_2\|$: indeed if π is a representation of involutive Banach algebras, then it has norm $\leqq 1$ ([4], p. 7) and since by the definition of the norm on $A_1 \hat{\otimes} A_2$ we have $\|a_1 \otimes a_2\| \leqq \|a_1\| \; \|a_2\|$, one obtains $\|\pi(a_1 \otimes a_2)\| \leqq \|a_1\| \; \|a_2\|$; conversely, if a representation of involutive algebras π of $A_1 \otimes A_2$ is given with $\|\pi(a_1 \otimes a_2)\| \leqq \|a_1\| \; \|a_2\|$, then with $x = \Sigma a_{1i} \otimes a_{2i}$ one has $\|\pi(x)\| \leqq \Sigma \|a_{1i}\| \; \|a_{2i}\|$, thus (by 2.4) we have $\|\pi(x)\| \leqq \|x\|$. Consequently π extends uniquely to a $*$-representation of $A_1 \hat{\otimes} A_2$. From this bijective correspondence we conclude <u>that our tensor product $\otimes^*$ coincides with the tensor product $\otimes^\nu$ introduced by GUICHARDET</u> [7]. Another possibly different tensor product has been introduced earlier by TAKESAKI and WULFSOHN [17]. If $\pi_i : A_i \longrightarrow B(H_i)$ are representations on Hilbert space, then $\pi_1 \otimes \pi_2 : A_1 \otimes A_2 \longrightarrow B(H_1 \otimes H_2)$ extends uniquely to a representation $\pi_1 \otimes^* \pi_2 : A_1 \otimes^* A_2 \longrightarrow B(H_1 \otimes H_2)$. Let $K = \bigcap\{\ker \pi_1 \otimes^* \pi_2 : \pi_i$ is a representation of A_i on a Hilbert space H_i, $i = 1,2\}$. Then $A_1 \otimes A_2 \cap K = \{0\}$ since $A_1 \otimes A_2$ is faithfully represented by $\pi_1 \otimes \pi_2$, if π_1 and π_2 are faithful. Let $A_1 \bigcirc^* A_2 = (A_1 \otimes^* A_2)/K$; this is the

Takesaki-Wulfsohn tensor product. In general, $K \neq \{0\}$, but one hardly knows when the two tensor products agree. What is known, however, is the fact, crucial for us, that $K = \{0\}$, i.e. $A_1 \otimes^* A_2 = A_1 \odot^* A_2$ if one of the C*-algebras A_1 or A_2 is of type I [7]. For the details concerning the nature of type I algebras we must refer to [3] and [4]; e.g. all postliminal (=GCR) C*-algebras are of ty e I (see [4],p.111). What is relevant for our purposes, in any case, is that certainly all commutative C*-algebras are of type I. Some of the conveniences of the tensor product $\odot^*$ which are perhaps not shared by $\otimes^*$, are these:

a) If $\pi_i : A_i \longrightarrow B(H_i)$, $i = 1,2$ are faithful representations, then $\pi_1 \odot^* \pi_2 : A_1 \odot^* A_2 \longrightarrow B(H_1 \otimes H_2)$ is faithful [17].

b) If $A_1 \longrightarrow A_2$ is an injective C*-morphism and B is any C*-algebra, then $A_1 \odot^* B \longrightarrow A_2 \odot^* B$ is injective. (Consequence of a).)

We observe the following:

7.8. <u>LEMMA</u>. Let I be a closed two-sided ideal of the C*-algebra A, so that $0 \longrightarrow I \longrightarrow A \xrightarrow{\pi} A/I \longrightarrow 0$ is an exact sequence. Let B be an arbitrary C*-algebra.

a) The sequence $I \otimes^* B \longrightarrow A \otimes^* B \longrightarrow (A/I) \otimes^* B \longrightarrow 0$ is exact.

b) If B is of type I, then also $0 \longrightarrow I \otimes^* B \longrightarrow A \otimes^* B$ is exact.

Proof. b) follows from the remark b) above and the fact that $-\bigcirc^* B = -\otimes^* B$ if B is of type I.

a) was proved by GUICHARDET [7].

<u>REMARK</u>. Condition a) says that the functor $-\otimes^* B$ is right exact (or cokernel preserving) in C* (see e.g. [13], p. 51), and b) asserts that it is in fact exact if B is of type I. Thus every type I C*-algebra is <u>flat</u>. (A C*-algebra B is <u>flat</u> iff $-\otimes^* B$ is exact.)

We collect the relevant information in the following form:

7.9. <u>FUNDAMENTAL THEOREM FOR C*-TENSOR PRODUCTS</u>. <u>Let</u> A_i, $i = 1,2$ <u>be</u> C*-<u>algebras and</u> $\otimes^*$ <u>the</u> C*-<u>tensor product defined in</u> 7.5.

(1) $A_1 \otimes A_2$ <u>may be identified with an involutive subalgebra of</u> $A_1 \otimes^* A_2$.

(2) <u>The function</u> $\otimes : A_1 \times A_2 \longrightarrow A_1 \otimes^* A_2$ <u>is bilinear and satisfies</u> $\|a_1 \otimes a_2\| \leq \|a_1\| \; \|a_2\|$. <u>If</u> $\varphi_i : A_i \longrightarrow A_3$, $i = 1,2$ <u>are</u> C*-<u>morphisms such that</u> $\varphi_1(A_1)$ <u>and</u> $\varphi_2(A_2)$ <u>commute elementwise, then there is a unique</u> C*-<u>morphism</u> $\varphi_3 : A_1 \otimes^* A_2 \longrightarrow A_3$ <u>with</u> $\varphi_3(a_1 \otimes a_2) = \varphi_1(a_1)\varphi_2(a_2)$. <u>If</u> A_i, $i = 1,2$ <u>has an identity, let</u> $\sigma_i : A_i \longrightarrow A_1 \otimes^* A_2$ <u>be the natural injections satisfying</u> $\sigma_1(a_1) = a_1 \otimes 1$, $\sigma_2(a_2) = 1 \otimes a_2$. <u>If</u> $\varphi_3 : A_1 \otimes^* A_2 \longrightarrow A_3$ <u>is a</u> C*-<u>morphism and one lets</u> $\varphi_i = \varphi_3 \sigma_i$, <u>then</u> $\varphi_1(a_1)\varphi_2(a_2) = \varphi_3(a_1 \otimes a_3)$ <u>and</u> $\varphi_1(A_1)$ <u>and</u> $\varphi_2(A_2)$ <u>commute elementwise. In the</u>

presence of identities the C*-tensor product is unique uniquely determined by these properties.

(3) For each C*-algebra A the function $-\otimes^* A$ is right exact, and in fact exact if A is of type I. In this case, if A_1 is a C*-subalgebra of A_2, then $A_1 \otimes^* A$ may be identified with a C*-subalgebra of $A_2 \otimes^* A$. If $A_1 \subset A_2$ and $B_1 \subset B_2$ are C*-algebras and A_1 and B_2 are flat, then $A_1 \otimes^* B_1$ may be considered as a subalgebra of $A_2 \otimes^* B_2$.

CONVENTION. It will be convenient to choose the following notation. If $A_1 \subseteq A_2$, $B_1 \subseteq B_2$ are C*-algebras, we will denote the image of $A_1 \otimes^* A_2 \longrightarrow B_1 \otimes^* B$ with $A_1 \underline{\otimes}^* A_2$. This notation of course is ambiguous, but the context will always make it clear which is the appropriate algebra containing $A_1 \underline{\otimes}^* A_2$. If in the course of a discussion all C*-algebras are flat, we may just ignore the underline. This holds in particular for the commutative case.

7.10. COROLLARY. Let $\pi: A \longrightarrow A/I$ and $\varphi: B \longrightarrow B/J$ be quotient morphisms in C*.

a) $\ker(\pi \otimes^* \varphi) = (I \underline{\otimes}^* B) + (A \underline{\otimes}^* J)$.

b) If A, B, I and J are flat, then

$$I \otimes^* J = (A \otimes^* J) \cap (I \otimes^* B).$$

Proof. a) has been proved by GUICHARDET [7]. (The methods of our proof of b) would yield a proof, too.)

b) Consider the diagram with exact rows and columns

$$\begin{array}{ccccccccc}
 & & 0 & & 0 & & 0 & & \\
 & & \downarrow & & \downarrow & & \downarrow & & \\
0 & \longrightarrow & I \otimes^* J & \longrightarrow & A \otimes^* J & \rightarrow & (A/I) \otimes^* J & \longrightarrow & 0 \\
 & & \downarrow & & \downarrow & & \downarrow & & \\
0 & \longrightarrow & I \otimes^* B & \longrightarrow & A \otimes^* B & \rightarrow & (A/I) \otimes^* B & \longrightarrow & 0 \\
 & & \downarrow & & \downarrow & & \downarrow & & \\
0 & \longrightarrow & I \otimes^*(B/J) & \rightarrow & A \otimes^*(B/J) & \rightarrow & (A/I)\otimes^*(B/J) & \rightarrow & 0 \\
 & & \downarrow & & \downarrow & & \downarrow & & \\
 & & 0 & & 0 & & 0 & &
\end{array}$$

Now we simply apply well-known arguments for exact categories to this diagram (see [13], p. 23 ff).

The applicability of categorical arguments in the previous proof is due to the following fact:

7.11. PROPOSITION. *The class* C^*_* *of all* C*-*morphisms* $\varphi : A \longrightarrow B$ *such that* $\varphi(A)$ *is an ideal in* B *is an exact category* ([13], p.18) *in which the monics are injective and the epics surjective.*

Proof. For any φ in C^*_* the image $\varphi(A)$ is closed [4]. If I is a closed ideal of a C*-algebra C and J is a closed ideal of I, then J is an ideal of C [4]. Hence $\psi\varphi \in C^*_*$ if $\psi : B \longrightarrow C$ is in C^*_*. Thus C^*_* is a subcategory of C* with the same objects as C*.

Further, every monic μ is injective (just as in C*), since $(\ker \mu \xrightarrow{c} A \xrightarrow{\mu} B) = (\ker \mu \xrightarrow{0} A \xrightarrow{\mu} B)$ implies $\ker \mu = 0$, and μ is obviously the kernel of $B \longrightarrow B/\mu(A)$. By Lorenz' result [12] every epic of C* is surjective; it is much simpler to see that an epic ϵ in C^*_* is surjective, since $(A \xrightarrow{\epsilon} B \xrightarrow{\text{quotient}} B/\epsilon(A)) = (A \xrightarrow{\epsilon} B \xrightarrow{0} B/\epsilon(A))$ implies $\epsilon(A) = B$. Clearly every

surjective morphism $\epsilon : A \longrightarrow B$ is the cokernel of
$\ker \epsilon \xrightarrow{\ c\ } A$. The exactness of $C^*_\#$ is clear since every
monic is a kernel, every epic a cokernel, and since every
morphism has an epic-monic factorization.

7.12. PROPOSITION. Let A be a C^*-algebra and A_i,
$i = 1, 2$, closed subalgebras of a C^*-algebra B. Then
$A \underline{\otimes}^* (A_1 + A_2)^- = (A \underline{\otimes}(A_1 + A_2)^-$ within $A \otimes^* B$. If the A_1
are ideals, then the closures may be omitted. If in this
case $A_1 \cap A_2 = \{0\}$, then all sums are direct.

Proof. We know that $A \otimes (A_1 + A_2)$ is dense in
$A \otimes (A_1 + A_2)^-$ which in turn is dense in $A \otimes^* (A_1 + A_2)^-$.
But $A \otimes (A_1 + A_2) = (A \otimes A_1) + (A \otimes A_2)$ which is dense in
$(A \underline{\otimes}^* A_1 + A \underline{\otimes}^* A_2)^-$. If the A_i are ideals we recall the
fact that in a C^*-algebra the sum of two closed ideal is
automatically closed. For the last remark remember 7.9
and apply $A \otimes^*$ - to the split exact sequence
$$0 \longrightarrow A_1 \longrightarrow A_1 \oplus A_2 \longrightarrow A_2 \longrightarrow 0.$$

7.13. PROPOSITION. Let A be a C^*-algebra with identity
and $m : A \otimes^* A \longrightarrow A$ an algebra over C^* in the sense of
3.1. Then there are C^*-morphisms $\varphi, \psi : A \longrightarrow A$ such that
    (1) $\varphi(a)\psi(b) = \psi(b)\varphi(a)$ for all $a, b \in A$,
    (2) $\varphi^2 = \varphi,\ \psi^2 = \psi$
    (3) $\varphi\psi = \psi\varphi$,
and (4) $m(a \otimes b) = \varphi(a)\psi(b)$ for all $a, b \in A$.
Conversely, if φ, ψ are given and satisfy (1), (2), (3),
then (4) defines an algebra $m : A \otimes^* A \longrightarrow A$ over C^*. If
m has an identity in the sense of 3.1, then $m(a \otimes b) = ab$,

i.e. m *is induced by the original multiplication. Thus an algebra with identity over* C* *in the sense of* 3.1 *is exactly a commutative* C*-*algebra with identity.*

Proof. As in 3.6.

8. *The cofunctor Spec for commutative* C*-*algebras*

For a C*-algebra A, let Prim A denote the space of its primitive ideals with the hull-kernel topology. If A is abelian, then Prim A and Spec A (see Section 6) may be identified. In the abelian situation, Spec is a cofunctor to the extent discussed in Section 6. In what sense Prim A is a cofunctor in the general case is not so clear even in the presence of an identity. The following remarks pertain to tensor products.

8.1 PROPOSITION. *Let* A_i *be* C*-*algebras and* $I_i \in \text{Prim } A_i$, $i = 1,2$. *Then* $(I_1 \underline{\otimes}^* A_2) + (A_1 \underline{\otimes}^* A_2) \in \text{Prim } A_1 \otimes^* A_2$.

Proof. After 7.10 we have to show that $(A_1/I_1) \otimes^* (A_2/J_2)$ is a primitive algebra. W.l.g. assume that $I_1 = 0$, $J_2 = 0$ and that A_1 and A_2 have faithful irreducible representations $\pi_i : A_1 \longrightarrow B(H_i)$, $i = 1, 2$ on Hilbert space. Then $\pi_1 \otimes^* \pi_2 : A_1 \otimes^* A_2 \longrightarrow B(H_1) \otimes^* B(H_2) \longrightarrow B(H_1 \otimes H_2)$ is a representation. We consider A_i as subalgebra of $B(H_i)$ and $B(H_1) \otimes B(H_2)$ as subalgebra of $B(H_1 \otimes H_2)$ so that $A_1 \otimes A_2$ is a subalgebra of $B(H_1 \otimes H_2)$. The irreducibility of A means that A_i is strongly dense in $B(H_i)$ (or, in other words, that the double commutator of A_i is all of $B(H_i)$

[3]). The double commutator of $A_1 \otimes A_2$ then is $B(H_1 \otimes H_2)$ (see [3]). Hence $A_1 \otimes A_2$ is irreducible on $H_1 \otimes H_2$ and thus so is its closure.

We will present the following fact without further discussion (see [7], [17], [18]; in [18] one seems to need the hypothesis that one of the factors is of type I).

8.2. For two C*-algebras A_i, $i = 1, 2$, one of which is of type I, the function $(I_1, I_2) \longrightarrow A_1 \underline{\otimes}^* I_2 + I_1 \underline{\otimes}^* A_2$: Prim $A_1 \otimes^* A_2$ is a homeomorphism onto a subspace of the range. It is a homeomorphism if one of the algebras A_i is of type I.

Let us now specialize our considerations directly to the category of commutative C*-algebras. In this case we may identify the primitive ideal space and the spectrum and in turn consider any one of these spaces as identified with the space of characters, i.e., multiplicative complex hermitean functionals with the weak star topology. For a pair of commutative C*-algebras A_i, $i = 1, 2$ we thus have a natural homeomorphism

$$h_{A_1, A_2} : \text{Spec } A_1 \times \text{Spec } A_2 \longrightarrow \text{Spec } A_1 \otimes^* A_2$$

which associates with a pair (f, g) of characters the character $A_1 \otimes^* A_2 \xrightarrow{f \otimes^* g} \mathbb{C} \otimes \mathbb{C} \xrightarrow{\cong} \mathbb{C}$, where the isomorphism is defined by the multiplication on $\mathbb{C}$. Conversely, if X_i, $i = 1, 2$ is a pair of locally compact spaces, then there is a natural isomorphism of commutative C*-algebras

$$t_{X_1,X_2}: C_o(X_1) \otimes^* C_o(X_2) \longrightarrow C_o(X_1 \times X_2)$$

which is given by $[t_{X_1,X_2}(f_1 \otimes f_2)](x_1,x_2) = f_1(x_1)f_2(x_2)$.

We may thus summarize

8.3. THEOREM. *Let* C^*_{prop} *be the category of commutative* C^**-algebras and proper morphisms,* C^*_a *the category of* C^**-algebras with identity (and identity preserving morphisms),* $\mathcal{L}$ *the category of locally compact spaces and proper maps,* Comp *the category of compact spaces.*

Then

$$\text{Spec}: (C^*_{prop}, \otimes^*) \longrightarrow (\mathcal{L}, \times)$$

$$\text{Spec}: (C^*_a, \otimes^*) \longrightarrow (\text{Comp}, \times)$$

$$C_o: \quad (\mathcal{L}, \times) \longrightarrow (C^*_{prop}, \otimes^*)$$

$$C: \quad (\text{Comp}, \times) \longrightarrow (C^*_a, \otimes^*)$$

are multiplicative cofunctors.

From the two theorems 6.6 and 8.3 above as point of departure, using the general principles outlined in Sections 3, 4, 5 it will be easy to formulate a complete duality theory for compact semigroups.

9. *Finite coproducts of commutative* C^**-algebras with identity*

9.1. DEFINITION. The full subcategory of all C^*-algebras with identity and identity preserving morphisms will be denoted with C^{*1}.

The following is a consequence of 7.8.

9.2. <u>PROPOSITION</u>. For two C*-algebras with identity A and B we have a diagram

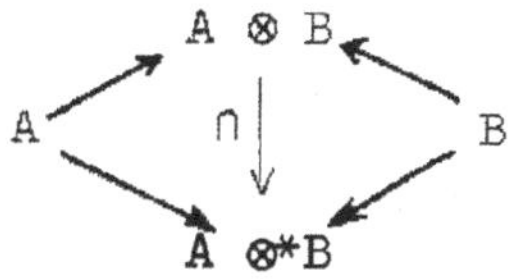

If $\varphi: A \longrightarrow C$ and $\psi: B \longrightarrow C$ are morphisms of C*-algebras with identity such that the images $\varphi(A)$ and $\psi(B)$ commute elementwise in C, then there is a unique morphism $\pi: A \otimes^* B \longrightarrow C$ such that $\pi s = \varphi$ and $\pi t = \psi$. In particular, in the category $C^{*1} \cap C^*_a$ the diagram

$$A \longrightarrow A \otimes^* B \longleftarrow C$$

is a coproduct.

It should be pointed out that the category of C*-algebras is both complete (allows limits) and cocomplete (allows colimits). The subcategory of commutative C*-algebras with identity is a complete and cocomplete subcategory. One therefore knows that the coproduct

$$A_1 \longrightarrow A_1 \coprod A_2 \longleftarrow A_2$$

of two commutative C*-algebras with identity exists. We have in fact identified this coproduct beforehand as the tensor product. This is in complete analogy to the case of (plain) commutative algebras with identity.

For C*-algebras without identity the situation is somewhat different, since in the absence of and identity there are no distinguished morphisms

$A_1 \longrightarrow A_1 \otimes^* A_2 \longleftarrow A_2$. The best surrogate is a continuous bilinear *-function $A_1 \times A_2 \xrightarrow{\otimes} A_1 \otimes^* A_2$ which has the property that for two C*-algebra morphisms $\varphi_i : A_i \longrightarrow B$ such that $\varphi_1(A_1)$ and $\varphi_2(A_2)$ commute elementwise, there is a unique C*-morphism $\varphi : A_1 \otimes^* A_2 \longrightarrow B$ with $\varphi(a_1 \otimes a_2) = \varphi_1(a_1)\varphi_2(a_2)$.

10. *C*-Cogebras*

10.1. *DEFINITION*. *A* C*-*cogebra* is a morphism of C*-algebras $c : A \longrightarrow A \otimes^* A$ for a C*-algebra A. *A* C*-*cogebra* *with* identity is a C*-cogebra in which A has an identity and c preserves identities. If c_1, c_2 are C*-cogebras, then *a morphism of* C*-*cogebras* [*with identity*] is a morphism $\varphi : A_1 \longrightarrow A_2$ of C*-algebras such that the diagram

$$\begin{array}{ccc} A_1 & \longrightarrow & A_1 \otimes^* A_1 \\ \varphi \downarrow & & \downarrow \varphi \otimes^* \varphi \\ A_2 & \longrightarrow & A_2 \otimes^* A_2 \end{array}$$

commutes.

A C*-*bigebra* is a diagram $A \xrightarrow{c} A \otimes^* A \xrightarrow{m} A$ of morphisms of C*-algebras satisfying the conditions of 5.1.

The concepts of *commutativity*, *cocommutativity*, *coidentity* are defined as in Sections 3, 4, 5. A C*-bigebra is idempotent if mc = identity of A.

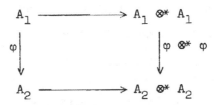

IMPORTANT REMARK: Note that every C*-cogebra $c: A \longrightarrow A \otimes^* A$ for which A is a commutative C*-algebra automatically defines a C*-bigebra

$$A \xrightarrow{c} A \otimes^* A \xrightarrow{m} A$$

with $m(a \otimes b) = ab$. In fact, if A has an identity, m is unique by 7.13. Thus the category co[C^*_a] of cogebras over the category C^*_a of commutative C*-algebras with identity and the category bi[C*] of bigebras over C* are naturally equal, and the category co[C^*_{prop}] of cogebras over the category C^*_{prop} of commutative C*-algebras and proper morphisms may be identified with a full subcategory of the category bi[C^*_{prop}] of bigebras over C^*_{prop}.

It may be observed that some authors may prefer to call a C*-bigebra what we call a C*-cogebra right away, since for any C*-cogebra $c: A \longrightarrow A \otimes^* A$ we obtain a diagram

$$A \xrightarrow{c} A \otimes^* A \xrightarrow{m} A$$

with a function $m: A \otimes^* A \longrightarrow A$ satisfying $m(a \otimes^* b) = ab$ as it is the case, e.g. in the category Inv. However, it must be pointed out that the existence of m is not guaranteed in the non-commutative case. We <u>do</u> get a morphism of involutive Banach spaces $\overline{m}: A \hat{\otimes} A \longrightarrow A$ such that $\overline{m}(a \hat{\otimes} b) = ab$. However, in the non-commutative case, $\overline{m}$ will not be generally a morphism of involutive Banach algebras. Hence it need not factor through $A \otimes^* A$. To repeat: In the commutative case the distinction between

C*-cogebra and C*-bigebra becomes virtually irrelevant (and certainly irrelevant if the cogebra also has an identity). Sometimes we will denote a C*-bigebra with A, if no confusion is possible.

We discuss the rudiments of ideal theory involved in the basics of bigebra theory.

10.2. PROPOSITION. *Let* $c:A \longrightarrow A \otimes^* A$ *be a* C*-*cogebra. If* I *is a closed ideal of* A *and* $\pi:A \longrightarrow A/I$ *the quotient morphism then the morphism* $A \xrightarrow{c} A \otimes^* A \xrightarrow{\pi \otimes^* \pi} (A/I) \otimes (A/I)$ *factors through* π *if and only if* $c(I) \subset I \underline{\otimes}^* A + A \underline{\otimes}^* I$. *In this case there is a unique morphism* $c_I:A/I \longrightarrow (A/I) \otimes^* (A/I)$ *such that the diagram*

$$\begin{array}{ccc} A & \longrightarrow & A \otimes^* A \\ \downarrow & & \downarrow \pi \otimes^* \pi \\ A/I & \longrightarrow & (A/I) \otimes^* (A/I) \end{array}$$

commutes.

Proof. The first assertion follows from $\ker(\pi \otimes^* \pi) = (I \underline{\otimes}^* A) + (A \underline{\otimes} I)$ (7.10). The second is immediate.

10.3. DEFINITION. If $c:A \longrightarrow A \otimes^* A$ is a c*-cogebra, then I is called a *bigebra kernel* if it is a closed ideal of A satisfying $c(I) \subset I \underline{\otimes}^* A + A \underline{\otimes}^* I$. The C*-cogebra $c_I:A/I \longrightarrow (A/I) \otimes^* (A/I)$, whose existence then is secured by proposition 10.2, is *a* C*-*cogebra quotient of* A.

10.4. PROPOSITION. *Let* $c: A \longrightarrow A \otimes^* A$ *be a* C^*-*cogebra. If* I *is a closed ideal of* A *and* $\pi: A \longrightarrow A/I$ *the quotient morphism, then the morphism*

$$A \xrightarrow{c} A \otimes^* A \longrightarrow (A/I) \otimes^* A$$

factors through π *if and only if* $c(I) \subset I \otimes A$. *In this case there is a unique morphism* $c_r: A/I \longrightarrow (A/I) \otimes^* A$ *such that the diagram*

$$\begin{array}{ccc} A & \xrightarrow{\quad c \quad} & A \otimes^* A \\ \pi \downarrow & & \downarrow \pi \otimes^* A \\ A/I & \xrightarrow[\quad c_r \quad]{} & (A/I) \otimes^* A \end{array}$$

commutes.

Proof. We note that $\ker \pi \otimes^* A = I \otimes^* A$ by 7.10.

10.5. DEFINITION. Let $c: A \longrightarrow A \otimes^* A$ be a C^*-cogebra. Then $I \subset A$ is called *a right bideal* if I is a closed ideal of A and $c(I) \subset I \otimes^* A$. A *left bideal* is defined analogously. Finally, $I \subset A$ is *a bideal* if I is closed ideal and is both a left and a right bideal.

In view of $A \otimes^* I \cap I \otimes^* A = I \otimes^* I$, if A is of type I (7.10) we now have

10.6. PROPOSITION. *Let* $c: A \longrightarrow A \otimes A$ *be a* C^*-*cogebra. Then* $I \subset A$ *is a bideal if and only if* I *is a closed ideal of* A *and* $c(I) \subseteq A \otimes^* I \cap I \otimes^* A$. *In this case there is a commutative diagram*

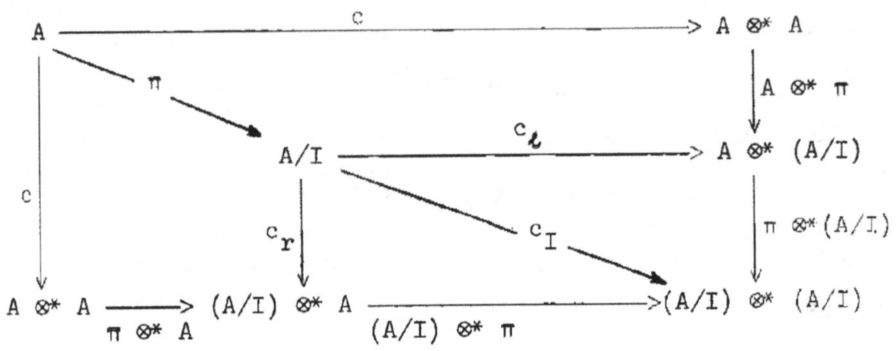

If A *is flat* (e.g. *of type* I), *then* I *is a bideal iff* $c(I) \subseteq I \otimes^* I$.

10.7. DEFINITION. A C*-cogebra is called

(a) *left cosimple,*

(b) *right cosimple,*

(c) *cosimple,*

(d) *colocal*

if

(a) it has no proper left bideals,

(b) it has not proper right bideals,

(c) it has no proper bideal,

(d) it has a uniquely determined maximal proper bideal.

REMARK. As to the terminology one might ask why the ostensibly important concept of a bigebra kernel is not classified terminologically in the hierarchy of bideals. This will become apparent when we go into the duality theory in whose framework it will become clear that in a suitable sense the bideals correspond to ideals in the dual objects. But as early as this point we may observe that if arrows are reversed in the pertinent diagrams in the preceding sections then the bigebra kernels correspond to the subalgebras and the various bideals to the various ideals. We admit that one might be tempted to introduce for a bigebra kernel the name of "newdeal", but we will not do so.

10.8. PROPOSITION. *In a* C^**-cogebra the set of bideals is closed under addition. The same holds for the set of left (respectively, right) bideals.*

Proof. Let $c: A \longrightarrow A \otimes^* A$ be a C^*-cogebra and I, J two left bideals. Then $c(I + J) = c(I) + c(J) \subseteq I \,\underline{\otimes}^*\, A + J \otimes^* A \subseteq (I + J) \,\underline{\otimes}^*\, A$, so $I + J$ is a left bideal.

10.9. PROPOSITION. a) *In a flat* C^**-cogebra the set of flat bideals is closed under intersection.*

b) *If* $c: A \longrightarrow A \otimes^* A$ *is a* C^**-cogebra and* I *(respectively,* J*) is a flat left (respectively, right) bideal, then* $I \cap J$ *is a bideal.*

Proof. b) $c(I \cap J) \subseteq c(I) \cap c(J) \subseteq (I \,\underline{\otimes}^*\, A) \cap (A \,\underline{\otimes}^*\, J) = I \,\underline{\otimes}^*\, J$.

a) follows from b).

We continue with a sequence of Lemmas all of which are part of the proof of a major theorem.

LEMMA a. The set of proper left bideals (right bideals) in a C*-cogebra with identity is inductive under inclusion.

Proof. Let F be an ascending tower of proper left bideals, and let I be the closure of the union. Then I is a closed proper ideal since the set of units is open in a Banach algebra with identity. Further $c(I) = c((\cup F)^{-} \subseteq c(\cup F)^{-}$ by the continuity of c. Since $c(\cup F = \cup\{c(J): J \in F\} \subseteq \cup\{J \underline{\otimes}^* A: J \in F\} \subseteq I \underline{\otimes}^* A$ since $J \underline{\otimes}^* A \subseteq I \underline{\otimes}^* A$ for all $J \in F$. Since $I \underline{\otimes}^* A$ is closed, $c(I) \subseteq I \underline{\otimes}^* A$ follows. Thus I is a left bideal and an upper bound of F in the set of left bideals.

LEMMA b. The set of proper bideals in a C*-cogebra with identity is inductive under inclusion.

Proof. If, in the proof of Lemma a, F is a tower of bideals, then I is also a right bideal by the same argument.

LEMMA c. In any C*-cogebra with identity there are maximal proper bideals, left bideals, right bideals.

Proof. Lemmas a, b, Zorn's lemma.

10.10. PROPOSITION. a) *Let* $c: A \longrightarrow A \otimes^* A$ *be a C*-cogebra with identity. Let I be a left bideal and J a right bideal. If* $A = I + J$, *then* $I = A$ *or* $J = A$.

b) *In particular, A is not a finite sum of proper bideals.*

Proof. We have $1 \otimes 1 = c(1) \in c(A) = c(I) + c(J) \subseteq A \otimes^* I + J \otimes^* A$, the latter is an ideal of $A \otimes^* A$, and since it contains the identity, it must equal $A \otimes^* A$. But $A \otimes^* I + J \otimes^* A \longrightarrow A \otimes^* A \longrightarrow (A/I) \otimes^* (A/J) \longrightarrow 0$ is exact by 7.10. Hence $A/I = 0$ or $A/J = 0$. This yields

LEMMA d. *If I, J are maximal bideals of a flat C^*-cogebra with identity then $I = J$, if I and J are proper.*

Proof. By 10.8. $I + J$ is a bideal, by 10.10, $I + J \neq A$. By maximality of I, $I = I + J$, so $J \subseteq I$. By symmetry, equality follows.

10.11. PROPOSITION. *Let $c: A \longrightarrow A \otimes^* A$ be a flat C^*-cogebra and let $I \subseteq A$ be a closed flat ideal of A, and I_ℓ and I_r left, respectively right, bideals. Then*

(a) $c^{-1}(I_\ell \otimes^* I)$ *is a left bideal,*

(b) $c^{-1}(I \otimes^* I_r)$ *is a right bideal,*

(c) *If* $d = (A \otimes^* c)c = (c \otimes^* A)c$, *then* $d^{-1}(I_\ell \otimes^* I \otimes^* I_r)$ *a bideal.*

(d) $c^{-1}(I_\ell \otimes I_r)$ *is a bideal.*

Proof. (a) Let $J = c^{-1}(I_\ell \otimes^* I)$. Then

$$\begin{array}{ccc} J & \xrightarrow{\ c'\ } & I_\ell \otimes^* I \\ \cap \downarrow & & \downarrow \cap \\ A & \xrightarrow[\ c\ ]{} & A \otimes^* A \end{array}$$

is a pull-back diagram in C^*, and

$0 \longrightarrow I_\ell \otimes^* I \longrightarrow A \otimes A \xrightarrow{k} K \longrightarrow 0$ is exact with a suitable K. Then J is the kernel of kc [13; p.15]. Forming tensor products with a flat object preserves kernels, so in the diagram

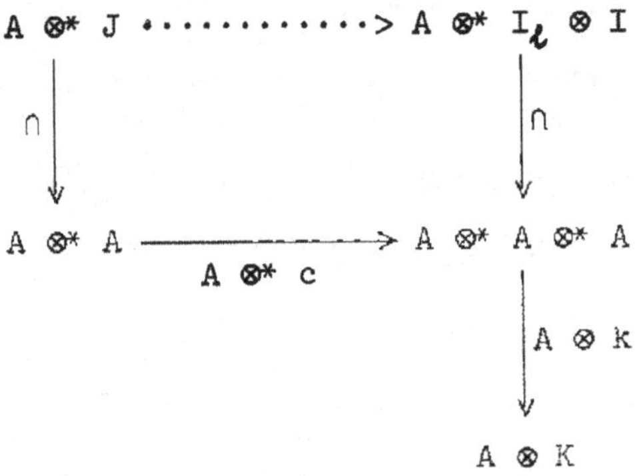

A ⊗* J is the kernel of A ⊗* ck = (A ⊗* c)(A ⊗ k). Hence the diagram can be completed to a pull back (loc. cit.), along ···>. Thus A ⊗* J = $(A \otimes^* c)^{-1}$ A ⊗* I_ℓ ⊗ I.

We have a commutative diagram

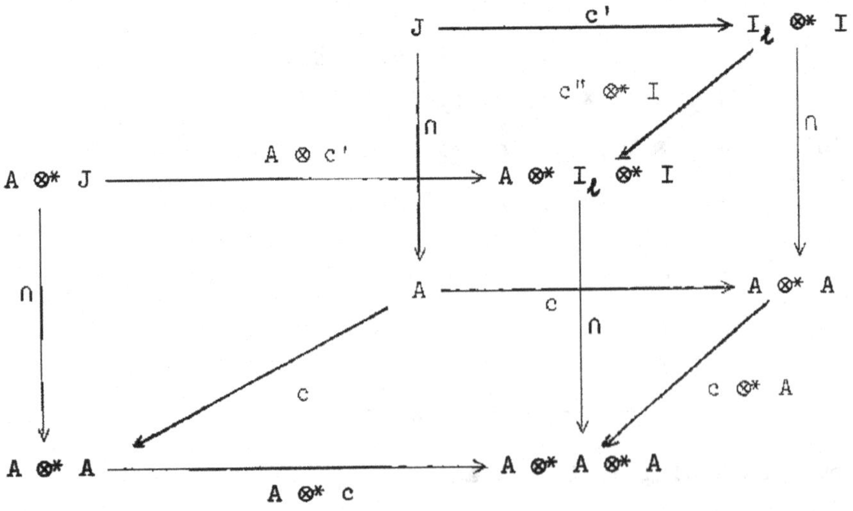

since I_ℓ is a left bideal. The front face being a pull back, there is a unique fill-in $J \longrightarrow A \otimes^* J$ commuting with the diagram. Thus $c(J) \subseteq A \otimes^* J$, i.e. J is a left bideal.

(b) is analogous to (a).

(c) Defining inverse images by pullbacks as in (a) and observing composition of pullbacks, $d^{-1}(I_\ell \otimes^* I \otimes^* I_r) = c^{-1}(I_\ell \otimes^* c^{-1}(I \otimes^* I_r))$; by (a) this is a left bideal. Similarly, one shows that it is a right bideal.

(d) follows from (a) and (b).

<u>LEMMA e</u>. Let $c:A \longrightarrow A \otimes A$ be a flat C*-cogebra with identity and I a maximal proper bideal. Let $d:A \longrightarrow A \otimes^* A \otimes^* A$ be defined by $d = (A \otimes^* c)c = (c \otimes^* A)c$.

(a) If J is any closed proper flat ideal of A containing I, then $I = d^{-1}(A \otimes J \otimes A) = d^{-1}(A \otimes A \otimes I) = d^{-1}(I \otimes A \otimes A)$.

(b) If I_ℓ, I_r are proper left, respectively right, bideals containing I, then $I = c^{-1}(I_\ell \otimes I_r)$.

Proof. (a) By 10.14, $d^{-1}(A \otimes^* J \otimes^* A)$ is a bideal. Since $d(I) = (A \otimes^* c)c(I) \subseteq (A \otimes^* c)(I \otimes^* I) \subseteq I \otimes^* I \otimes^* I \subseteq A \otimes^* J \otimes^* A$ we have $I \subseteq d^{-1}(A \otimes^* J \otimes^* A)$. Since $0 \longrightarrow A \otimes^* J \otimes^* A \longrightarrow A \otimes^* A \otimes^* A \longrightarrow A \otimes^* (A/J) \otimes^* A \longrightarrow 0$ is exact and the third morphism is not zero, $A \otimes^* J \otimes^* A$ and thus $d^{-1}(A \otimes^* J \otimes^* A)$ are proper ideals. By the maximality of I the assertion follows, and the remaining cases are similar.

(b) By 10.10 $c^{-1}(I_\ell \otimes^* I_r)$ is a bideal. Since $c(I) \subseteq I \otimes^* I \subseteq I_\ell \otimes^* I_r$ we have $I \subseteq c^{-1}(I_\ell \otimes^* I_r)$. Since $I_\ell \otimes^* I_r \subseteq I_\ell \otimes^* A = \ker(A \otimes^* A \longrightarrow A/I_\ell \otimes^* A)$, then $I_\ell \otimes I_r$ and $d^{-1}(I_\ell \otimes I_r)$ are proper ideals. So, by maximality of I, the assertion follows.

<u>LEMMA f</u>. Let $c:A \longrightarrow A \otimes^* A$ be a flat C*-cogebra with identity and I a maximal bideal (which is unique by Lemma d). Then $c_I:A/I \longrightarrow (A/I) \otimes^* (A/I)$ is a cosimple C*-bigebra.

Proof. Suppose J is a closed ideal of A containing I such that J/I is a bideal of c_I. Then there is a commutative diagram

$$
\begin{array}{ccccc}
0 & & & & 0 \\
\downarrow & & & & \downarrow \\
J & & & & \begin{array}{c} I \otimes^* A \otimes^* A \\ +A \otimes^* J \otimes^* A \\ +A \otimes^* A \otimes^* I \end{array} \\
\downarrow & & & & \downarrow \\
A & \xrightarrow{\;c\;} & A \otimes^* A & \xrightarrow{\;A \otimes^* c\;} & A \otimes^* A \otimes^* A \\
\downarrow{\scriptstyle \pi_J} & & \downarrow{\scriptstyle \pi_I \otimes^* \pi_J} & & \downarrow{\scriptstyle \pi_I \otimes^* \pi_J \otimes^* \pi_I} \\
A/J & \xrightarrow[c_\ell]{} & A/I \otimes^* A/J & \xrightarrow[(A/I) \otimes^* c_r]{} & A/I \otimes^* A/J \otimes^* A/I \\
\downarrow & & & & \downarrow \\
0 & & & & 0
\end{array}
$$

in which $c_\ell:A/J \longrightarrow (A/I) \otimes^* (A/J)$ and

$c_r:A/J \longrightarrow (A/J) \otimes^* (A/I)$

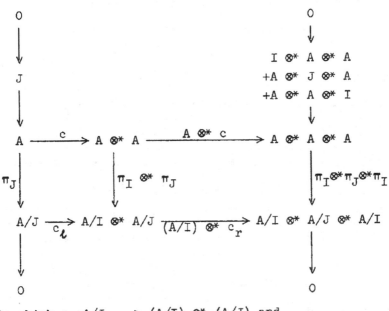

are induced by c_I due to the fact that J/I is a bideal of A/I (see 10.4 ff). The vertical sequences are exact. Denote $(A \otimes^* c)c$ with d as in 10.10. Then

$$d^{-1}(I \otimes^* A \otimes^* A + A \otimes^* J \otimes^* A + A \otimes^* A \otimes^* I) = I$$

by Lemma e, and so I is the kernel of $(\pi_I \otimes^* \pi_J \otimes^* \pi_I)d = ((A/I) \otimes^* c_r)c_\ell \pi_J$. Hence it contains $\ker \pi_J = J$. Thus $J = I$.

We have now proved the following theorem, which may be a little surprising in view of the fact that for algebras the property of being local (i.e. of having a unique maximal proper ideal) is quite restrictive.

10.12. THE COLOCALITY THEOREM FOR C*-COGEBRAS. *Any C*-cogebra $c: A \longrightarrow A \otimes^* A$ with identity is colocal. If A is flat, then the quotient cogebra modulo the unique maximal bideal I is cosimple. If I_r (respectively I_ℓ) is a maximal proper right (respectively left) bideal, then $I = c^{-1}(I_\ell \otimes^* I_r)$.*

10.13. REMARK. The colocality theorem prevails for cogebras with identity over the category of algebras over $\mathbb{C}$ and over the category of involutive algebras over $\mathbb{C}$.

The proofs are the same as for C*-algebras (they are, in fact, a little simpler).

This raises the question of a possible adjunction of the identity to a C*-cogebra.

10.14. Let $c: A \longrightarrow A \otimes A$ be a C*-cogebra. Let $\tilde{A} = \mathbb{C} \oplus A$ be the unique C*-algebra obtained by adjoining an identity to A [4]. Define $\tilde{c}: \tilde{A} \longrightarrow \tilde{A} \otimes \tilde{A} = \mathbb{C} \oplus A \oplus A \oplus A \otimes^* A$ (where $\mathbb{C} \otimes \mathbb{C}$ is identified with $\mathbb{C}$, $\mathbb{C} \otimes A$ with A and $A \otimes \mathbb{C}$ with A) by $\tilde{c}(\lambda + a) = \lambda \oplus 0 \oplus 0 \oplus c(a)$. Then $\tilde{c}$ is a C*-cogebra with identity, containing A as a bideal such that the quotient algebra is the C*-cogebra $\mathbb{C}$, and A is the maximal bideal of $\tilde{A}$. If A is commutative, then $\tilde{c}$ is a C*-bigebra with identity.

Proof. $\tilde{c}((\lambda + a)(\mu + b)) = \tilde{c}(\lambda\mu + (\lambda b + \mu a + ab)) = \lambda\mu \oplus \lambda c(b) + \mu c(a) + c(a)c(b) = (\lambda \oplus c(a))(\mu \oplus c(b)) = \tilde{c}(\lambda + a)\tilde{c}(\mu + b)$; thus $\tilde{c}$ is obviously a morphism of C*-algebras with identity (preserving identities). Furthermore, $(\tilde{c} \otimes \tilde{A})\tilde{c}(\lambda + a) = \lambda \oplus (c \otimes^* A)c(a)$ and $(\tilde{A} \otimes^* \tilde{c})\tilde{c}(\lambda + a) = \lambda \oplus (A \otimes^* c)c(a)$. Thus $\tilde{c}$ is a cogebra. Since all commutative C*-cogebras with identity are C*-cogebras in a unique fashion, then $\tilde{c}$ is in fact a C*-bigebra, if A is commutative. Clearly A is a bideal, and since the quotient bigebra $\tilde{A}/A$ is isomorphic to $\mathbb{C}$, the maximal bideal of $\tilde{A}$ is exactly A.

The following proposition rounds this topic off by discussing the possibility of adjoining a coidentity.

10.15. Let $c: A \longrightarrow A \otimes^* A$ be a C*-cogebra. Let A^1 be the C*-algebra product $\mathbb{C} \times A$. Define $c^1: A^1 \longrightarrow A^1 \otimes^* A^1$ by $c^1(\lambda \oplus a) = \lambda \oplus a \oplus a \oplus c(a) \in \mathbb{C} \times A \times A \times c(A)$ (with similar identifications as in 10.14). Let $\epsilon: A^1 \longrightarrow \mathbb{C}$ be the first projection. Then c^1 is a C*-cogebra with coidentity ϵ.

Proof. $c^1((\lambda \oplus a)(\mu \oplus b)) = c^1(\lambda\mu \oplus ab) =$ $\lambda\mu \oplus ab \oplus ab \oplus c(ab)) = (\lambda \oplus a \oplus a \oplus c(a))(\mu \oplus b \oplus b \oplus c(b))$ so c^1 is a C*-morphism. On one hand we have $(c^1 \otimes^* A^1)c^1(\lambda \oplus a)$ $= (c^1 \otimes^* A)((1 \otimes^* \lambda) \oplus (1 \otimes^* a) \oplus (a \otimes^* 1) \oplus (a \otimes^* a)) =$ $1 \otimes^* 1 \otimes^* (\lambda + a) + c(a) \otimes^* 1 + (c \otimes^* A)c(a) \in A^1 \otimes A^1 \otimes A^1$, on the other $(c^1 \otimes^* A^1)c^1(\lambda \oplus a)(A \otimes^* c^1)c^1(\lambda \oplus a) =$ $1 \otimes^* 1 \otimes^* \lambda \oplus 1 \otimes 1 \otimes a \oplus c(a) \otimes^* 1 \oplus (A \otimes^* c)c(a)$. Thus c^1 is coassociative. Finally $(\epsilon \otimes^* A)c(\lambda \oplus a) =$ $(\epsilon \otimes A)((1 \otimes^* \lambda) \oplus (1 \otimes^* a) \oplus (a \otimes^* 1) \oplus c(a)) =$ $1 \otimes^* \lambda \oplus 1 \otimes^* a = 1 \otimes^* (\lambda \oplus a)$. Similarly $(A \otimes^* \epsilon)c(\lambda \oplus a)$ $= (\lambda \oplus a) \otimes^* 1$.

Let us now make the important observation that with a pair consisting of a C*-cogebra and a C*-algebra there is always associated a topological semigroup.

If A_1, A_2 are C*-algebras, let as usual $C^*(A_1,A_2)$ denote the set of C*-algebra morphisms. There are many topologies on this set; let us use the topology of pointwise convergence induced from $C^*(A_1,A_2) \subseteq A_2^{A_1}$.

Suppose now that $c: A_1 \longrightarrow A_1 \otimes^* A_1$ is a C*-cogebra and let A_2 be a commutative C*-algebra. Let $m: A_2 \otimes^* A_2 \longrightarrow A_2$ be the C*-morphism defined by multiplication on A_2. (It suffices to assume that m is an algebra over C*.) Take $\varphi, \psi \in C^*(A_1,A_2)$ and define $\varphi\psi \in C^*(A_1,A_2)$ as the following composition:

$$A_1 \xrightarrow{c} A_1 \otimes^* A_2 \xrightarrow{\varphi \otimes^* \psi} A_2 \otimes^* A_2 \xrightarrow{m} A_2 .$$

For φ,ψ,ρ ∈ C*(A_1,A_2) one has a commutative diagram

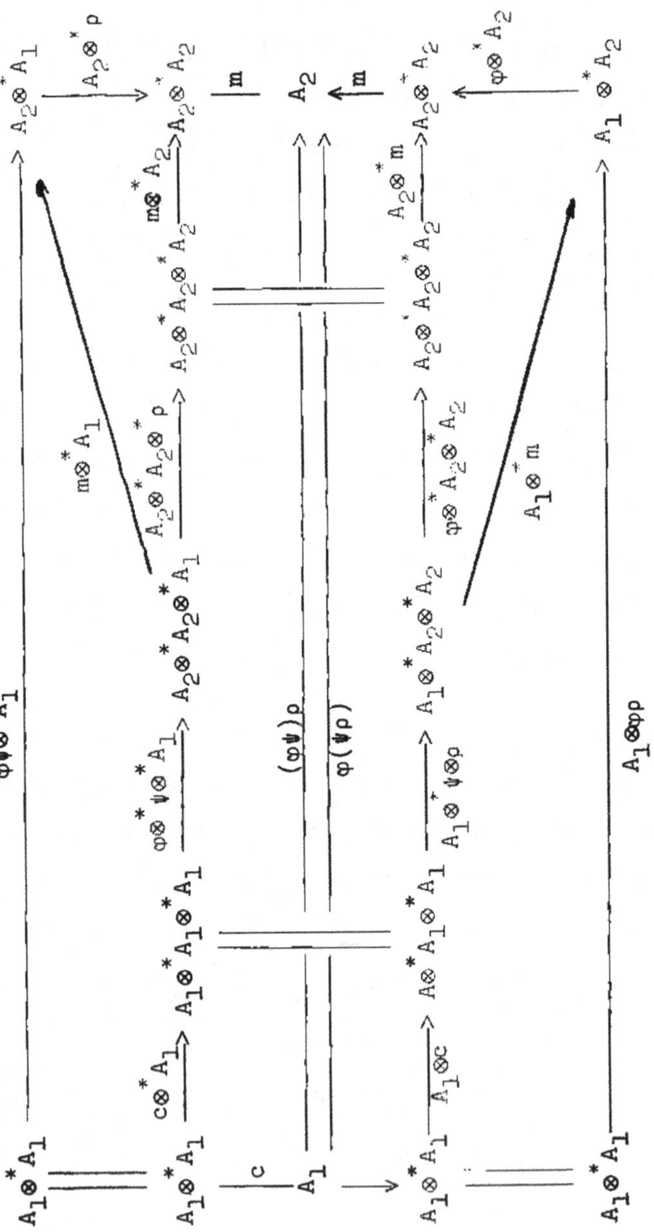

Hence $C^*(A_1,A_2)$ is a semigroup. The function $(\varphi,\psi) \longrightarrow \varphi \otimes^* \psi$ from $C^*(A_1,A_2)^2$ into $C^*(A_1 \otimes^* A_1, A_2 \otimes^* A_2)$ is continuous, for let $(\varphi,\psi) = \lim(\varphi_i,\psi_i)$ in $A_1 \times A_2$. Then $\lim(\varphi_i \otimes^* \psi_i)(\sum_{j=1}^{m} a_j \otimes^* b_j) = \sum_{j=1}^{m} \lim(\varphi_i(a_j) \otimes^* \varphi_i(b_j)) = (\varphi \otimes^* \psi)(\sum_{j=1}^{m} a_j \otimes^* b_j)$. Let now $b \in A_1 \otimes^* A_1$; if $\epsilon > 0$ is any positive number, then there is a $c \in A_1 \otimes A_1$ with $\|b - c\| < \epsilon$. We find an index i_o such that $i > i_o$ implies $\|(\varphi_i \otimes^* \psi_i)(c) - (\varphi \otimes^* \psi)(c)\| < \epsilon$. The morphisms $\varphi \otimes^* \psi$, $\varphi_i \otimes^* \psi_i$ as morphisms of C*-algebras have norms not exceeding 1. For $i > i_o$ we now have

$$\begin{aligned}
&\|(\varphi \otimes^* \psi)(b) - (\varphi_i \otimes^* \psi_i)(b)\| \\
\leqq\ &\|(\varphi \otimes^* \psi)(b) - (\varphi \otimes^* \psi)(c)\| \\
+\ &\|(\varphi \otimes^* \psi)(c) - (\varphi_i \otimes^* \psi_i)(c)\| \\
+\ &\|(\varphi_i \otimes^* \psi_i)(c) - (\varphi_i \otimes^* \psi_i)(b)\| \\
\leqq\ &\|b - c\| + \epsilon + \|b - c\| \leqq 3\,\epsilon.
\end{aligned}$$

Thus $o \leqq \overline{\lim}\, \|(\varphi \otimes^* \psi)(b) - (\varphi_i \otimes^* \psi_i)(b)\| \leqq 3\,\epsilon$ for arbitrary $\epsilon > 0$. Thus $\lim(\varphi_i \otimes^* \psi_i)(b) = (\varphi \otimes^* \psi)(b)$ for all $b \in A_1 \otimes^* A_1$. Hence $C^*(A_1,A_2)$ is a topological semigroup.

Let U_i be the unit balls in A_i, $i = 1, 2$. Then $\varphi \in C^*(A_1,A_2)$ maps $\varphi(U_1)$ into U_2 and $\varphi \longrightarrow \varphi|U_1 : C^*(A_1,A_2) \longrightarrow U_2^{U_1}$ is a homeomorphism onto its image. If $\varphi_i|U_1$ converges to $\psi: U_1 \longrightarrow U_2$ then ψ extends uniquely to a

$C*$-morphism $\varphi: A_1 \longrightarrow A_2$, and $\lim \varphi_i = \varphi$. Thus the image of $C*(A_1, A_2)$ in $U_2^{U_1}$ is closed. This implies that $C*(A_1, A_2)$ is compact if $\dim A_2 < \infty$. Note that $C*(A_1, A_2)$ is a semigroup with zero, namely the zero morphism.

Suppose that $\epsilon: A_1 \longrightarrow \mathcal{C}$ is a coidentity for A_1 and $1: \mathcal{C} \longrightarrow A_2$ is an identity for m. Then the diagram

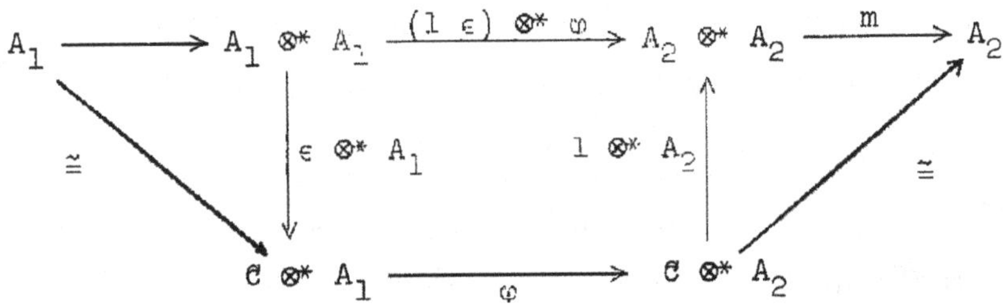

(and a dual one) shows that 1ϵ is an identity for $C*(A_1, A_2)$.

We have thus established the following theorem:

10.16.   THEOREM.   Let $c: A_1 \longrightarrow A_1 \otimes* A_1$ be a $C*$-cogebra and $m: A_2 \otimes* A_2 \longrightarrow A_2$ an algebra over $C*$ (i.e. A_2 is a $C*$-algebra and m an associative $C*$-morphism; if A_2 is commutative, then the multiplication of A_2 induces such an m). Let $C*(A_1, A_2)$ be the space of all $C*$-morphisms $A_1 \longrightarrow A_2$ relative to the topology of pointwise convergence. Then $C*(A_1, A_2)$ is a topological semigroup with zero via the diagram

$$
\begin{array}{ccc}
A_1 & \dashrightarrow & A_2 \\
{\scriptstyle c}\downarrow & & \uparrow{\scriptstyle m} \\
A_1 \otimes* A_1 & \xrightarrow{\;\varphi \otimes* \psi\;} & A_2 \otimes* A_2
\end{array}
$$

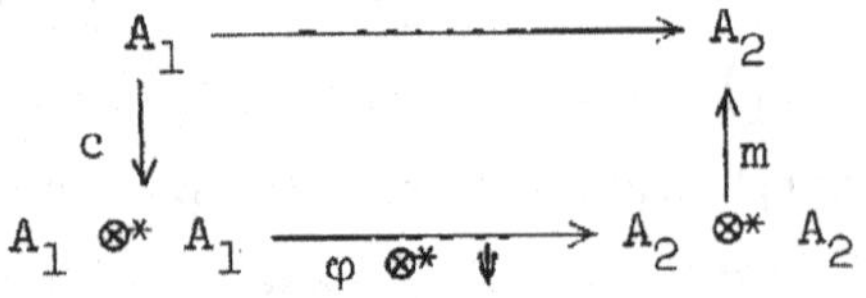

If $\epsilon: A_1 \longrightarrow \mathbb{C}$ *is a coidentity of* c, *and* $1: \mathbb{C} \longrightarrow A_2$ *an identity of* m, *then* $1\,\epsilon$ *is an identity for* $C^*(A_1, A_2)$. *If* $\dim A_2 < \infty$, *then* $C^*(A_1, A_2)$ *is compact.*

Let us look at some examples. The first is one which will preoccupy us for most of the remainder of this treatise. Let $c: A \longrightarrow A \otimes^* A$ be a C*-cogebra and consider the natural unique C*-algebra $\mathbb{C}$ with identity. Then $C^*(A, \mathbb{C})$ is a compact topological semigroup. If A is commutative, then $C^*(A, \mathbb{C}) = \operatorname{Spec} A \cup \{0\}$; if A has an identity, then 0 is isolated, otherwise not. If A is not commutative, let I be the closed ideal generated by all elements $ab - ba$. Then A/I is a commutative C*-algebra (the largest commutative quotient of A) and $C^*(A, \mathbb{C}) \cong C^*(A/I, \mathbb{C})$ as topological semigroups.

The second example is in some sense dual; consider the unique C*-cogebra $c: \mathbb{C} \longrightarrow \mathbb{C} \otimes^* \mathbb{C}$ which is an isomorphism given by $c(a) = 1 \otimes^* a$ and let $m: A \otimes^* A \longrightarrow A$ be a commutative C*-algebra. Now $C^*(\mathbb{C}, A)$ is in bijective correspondence with the set E(A) of idempotents of A; if $\varphi \in C^*(\mathbb{C}, A)$, then $\varphi(1) \in E(A)$ and if $e \in E(A)$, then $a \longmapsto a.e$ is in $C^*(\mathbb{C}, A)$, and the semigroup $C^*(\mathbb{C}, A)$ is the semilattice of idempotents in A.

There are further semigroups which are functorially associated with C*-bigebras. The following definition introduces a fundamentally important concept.

10.17. <u>DEFINITION</u>. Let $c: A \longrightarrow A \otimes^* A$ be a C*-cogebra. An element $a \in A$ is called *primitive*, if $c(a) = a \otimes a$. If c has an identity 1, then 1 is primitive. Clearly, 0 is always primitive. The set of all primitive elements will be called P(A).

10.18. <u>THEOREM</u>. *Let* $c: A \longrightarrow A \otimes^* A$ *be a* C*-*cogebra. Then* P(A) *is a subsemigroup of the multiplicative semigroup of* A *with the following properties:*

(i) P(A) *is closed under the involution* (i.e. *is an involutive semigroup*).

(ii) P(A) *is linearly independent.*

(iii) P(A) *is contained in the unit ball of* A.

(iv) P(A) *is closed.*

<u>REMARK</u>. We will later (in 11.11) prove the following amplification of (iii):

(iii') *If* $a \in P(A) \setminus \{0\}$, *then* $\|a\| = 1$, *provided* A *has a coidentity.*

(In the absence of a coidentity this is not generally correct.)

Proof. The semigroup property, (i) and (iv) are straightforward.

(ii) Let us show linear independence. Suppose that $\{p_o, \ldots, p_n\} \subseteq P(A) \setminus \{0\}$ is a minimal non-empty linearly dependent set. Then $p_o = \sum_{i=1}^{n} c_i p_i$ for suitable $c_i \in \mathbb{C}$. Now $c(p_o) = p_o \otimes p_o$, whence $\sum_{i=1}^{n} c_i p_i \otimes p_i = \sum_{i=1}^{n} c_i\, c(p_i) =$

$c(p_o) = p_o \otimes p_o = \sum_{i,k=1}^{n} c_i c_k p_i \otimes p_k$. By minimality, $\{p_1, \ldots, p_n\}$ is linearly independent. Hence $c_i = c_i$ for all i and $c_i c_k = 0$ for $i \neq k$. Thus $c_i = 0$ or $c_i = 1$. If $c_j = 1$, then $c_i = 0$ for $i \neq j$, and $p_o = p_j$, in which case, by minimality $j = 1$ and $\{p_o, p_i\} = \{p_o\}$, which is impossible. Thus $c_i = 0$ for all i yielding $p_o = 0$, which is likewise a contradiction.

(iii) Finally, observe that $\|a \otimes a\| = \|a\|^2$. (For let $\tilde{A}$ be the C*-algebra obtained by adjoining an identity (10.14). Then in $\tilde{A}$, $\|a^*a \otimes a^*a\| = \|(a^*a \otimes 1)(1 \otimes a^*a)\|$ $= \|a^*a \otimes 1\| \ \|1 \otimes a^*a\| = \|a^*a\|^2 = \|a\|^4$ since $\|a \otimes 1\|$ $= \|a\|$, as $a \longmapsto a \otimes 1 : A \longrightarrow A \otimes^* A$ defines an injection of C*-algebras. Thus, indeed, $\|a \otimes a\|^2 = \|(a \otimes a)^*(a \otimes a)\|$ $= \|a\|^4$.)

Hence $\|a\|^2 = \|a \otimes a\| = \|c(a)\| \leq \|a\|$, whence $\|a\| \leq 1$.

<u>REMARK</u>. We observe later, that as an immediate consequence of 10.18 (ii), the linear span A_p' of P(A) is the so-called reduced semigroup algebra $\mathbb{C}_{red}[P(A)]$ (= factor algebra of the semigroup algebra of P(A) modulo the algebra ideal generated by the zero of P(A)).

11. <u>The duality of compact semigroups</u>

At this stage we are more than ready to formulate an appropriate duality theory for compact semigroups. All material necessary has been provided. We recall and introduce some notation:

If $\mathfrak{A}$ is any category with a multiplication, then the category of cogebras (respectively, bigebras) over $\mathfrak{A}$ is denoted with co[$\mathfrak{A}$] (respectively, bi[$\mathfrak{A}$]). We recall: C^*_{prop} = Category of commutative C*-algebras and proper morphisms,

C^*_a = Category of commutative C*-algebras with identity and identity preserving morphisms.

$\mathcal{LC}$ = Category of locally compact proper topological semigroups and proper morphisms. [A topological semigroup is proper, if multiplication is a proper map, i.e. is closed and assigns to each point in the image a compact inverse image.]

$\mathcal{C}$ = Category of compact topological semigroups.

If $c:A \longrightarrow A \otimes^* A$ is a commutative C*-cogebra (and then a C*-bigebra) over C^*_{prop} (which means that c is a proper algebra morphism), then Spec $A \otimes^* A$ may be canonically identified with Spec A x Spec A and Spec c:Spec A x Spec A $\longrightarrow$ Spec A is a locally compact proper semigroup. If $m:A \otimes A \longrightarrow A$ is induced by the algebra multiplication, then Spec m:Spec A $\longrightarrow$ Spec A x Spec A is just the diagonal map. It is clear that in this fashion we define a cofunctor Spec: co[C^*_{prop}] $\longrightarrow \mathcal{LC}$. Conversely, let S be a locally compact proper semigroup with multiplication $m:S \times S \longrightarrow S$. The C*-algebra $C_o(S \times S)$ may be canonically identified with $C_o(S) \otimes^* C_o(S)$ and $C_o(m):C_o(S) \longrightarrow C_o(S) \otimes^* C_o(S)$ is a bigebra over C^*_{prop}. We thus have a cofunctor $C_o:\mathcal{LC} \longrightarrow co[C^*_{prop}]$. Remember that $co[C^*_{prop}]$ is a full subcategory of $bi[C^*_{prop}]$, and that $co[C^*_a] = bi[C^*_a]$.

11.1. DUALITY THEOREM FOR COMPACT AND LOCALLY COMPACT PROPER SEMIGROUPS. I) The cofunctors $C_o:\mathcal{LC} \longrightarrow co[C^*_{prop}]$ and $Spec:co[C^*_{prop}] \longrightarrow \mathcal{LC}$ are a duality, and the category of commutative proper C*-cogebras is dual to the category of locally compact proper semigroups.

II) The cofunctors $C:\mathcal{C} \longrightarrow bi[C^*_a]$ and $Spec:bi[C^*_a] \longrightarrow \mathcal{C}$ are a duality, which is compatible with the duality in I (see 1.3). The category of commutative C*-bigebras with identity is dual to the category of compact semigroups.

Let us test this duality theory by giving a brief run-down of the translation of elementary properties and concepts in the theory of compact semigroups in terms of this duality.

11.2. PROPOSITION. Let $c:A \longrightarrow A \otimes^* A$ be a proper commutative C*-cogebra, and let I be a closed ideal of A. Then the quotient morphism $\pi:A \longrightarrow A/I$ is proper, and Spec A/I may be identified with a closed subspace of Spec A via Spec π, and the following statements hold:

i) Spec A/I _is a subsemigroup if and only if_ I _is a bigebra kernel_, i.e. $c(I) \subseteq I \otimes^* A + A \otimes^* I$.

ii) Spec A/I _is a left ideal in the semigroup_ Spec A _if and only if_ I _is a left bideal_, _i.e._ $c(I) \subseteq A \otimes^* I$.

iii) Spec A/I _is an ideal in the semigroup_ Spec A _if and only if_ I _is a bideal_, i.e. $c(I) \subseteq I \otimes^* I$.

Proof. These statements now follow immediately from Section 10 by duality.

The colocality theorem for C*-cogebras with identity now yields the following result:

11.3. _THEOREM_. _Every compact semigroup has a unique minimal ideal which is simple_.

This is, of course, well known for compact semigroups and may be proved directly, but it appears in the present context as a special case of a much more general theorem, namely the colocality theorem for C*-cogebras 10.12.

We note that the additive semigroup $S = \{0,1,2,\ldots\}$ of natural numbers is proper. Its dual $C_o(S)$ therefore is a commutative proper C*-bigebra.

Since S has no minimal ideal, $C_o(S)$ is not colocal. We should note that S has an identity, whence $C_o(S)$ has a coidentity. Since S is commutative, $C_o(S)$ is cocommutative. Thus the colocality theorem pertains strictly to C*-cogebras with identity.

11.4. PROPOSITION. *Let* $c:A \longrightarrow A \otimes^* A$ *be a commutative proper* C*-*cogebra. Then the following statements hold:*

i) A *has an identity if and only if* Spec A *is compact.*

ii) A *has a coidentity if and only if* Spec A *has an identity.*

iii) A *is cocommutative if and only if* Spec A *is commutative.*

Recall that a C*-bigebra $A \longrightarrow A \otimes^* A \longrightarrow A$ is idempotent if $mc = 1_A$. By abuse of notation we call a C*-cogebra $c:A \longrightarrow A \otimes^* A$ over a commutative C*-algebra idempotent if the associated C*-bigebra (where m is induced by algebra multiplication) is idempotent.

11.5. PROPOSITION. a) *Let* $c:A \longrightarrow A \otimes^* A$ *be an object in* $co[C^*_{prop}]$. *Then* c *is idempotent if and only if* Spec A *is an idempotent semigroup. Further,* c *is idempotent and cocommutative if and only if* Spec A *is a semilattice* (i.e., *a commutative idempotent semigroup*).

b) *The category of idempotent cocommutative* C*-*bigebras with identity is dual to the category of compact semilattices.*

11.6. PROPOSITION. *Let* $c:A \longrightarrow A \otimes^* A$ *be a proper commutative* C*-*bigebra. Then the following statements are equivalent:*

a) Spec A *is a compact group.*

b) Spec A *is a group.*

c) _A is both right and left cosimple and has an identity._

d) _A has an identity, a left coidentity and is right cosimple._

e) _A has an identity, coidentity and is cosimple._

Proof. a) ==> b) is trivial. a) <==> c) <==> d) <==> e) is trivial by duality if one assumes the structure of the minimal ideal of a compact semigroup is known (see [11], p. 57).

b) ==> a). A locally compact but not compact topological group is not a proper semigroup, since multiplication is not a proper map.

11.7. COROLLARY. _The categories of compact groups and commutative cosimple C*-bigebra with identity and coidentity are dual to each other._

One might observe that in a commutative C*-bigebra with identity the comultiplication describes the algebraic properties of the dual semigroup and the multiplication its topological ones. The following comments illustrate this remark.

Let A be a commutative C*-algebra with identity; define E(A) to be the set of idempotents of A with the addition $e \oplus f = e + f - ef$ and the multiplication induced from A. Then E(A) is a closed subset of A and a Boolean algebra with identity. If A is identified with C(Spec A), then E(A) consists exactly of the locally constant functions Spec A $\longrightarrow \{0,1\}$. Let Spec E(A) be the set of all morphisms of Boolean algebras into the 2-element Boolean algebra GF(2) with the topology induced from $GF(2)^{E(A)}$. Then for each $f \in$ Spec A the function $f|E(A)$ is in Spec E(A). Thus there is a subjective continuous function λ:Spec A $\longrightarrow$ Spec E(A), and the inverse images $\lambda^{-1}(f)$ of single points are exactly the connected components of A.

We thus have the following remarks:

(i) Spec A _is connected if and only if_ $E(A) \cong GF(2)$ _(the two element algebra)_.

(ii) Spec A _is totally disconnected if and only if_ λ _is a homeomorphism if and only if for any two characters_ f_i of A_1, $i = 0, 1$, _there are idempotents_ e_i _of_ A _such that_ $f_o(e_o) = 0$ _and_ $f_1(e_1) = 1$.

This implies the following

11.8. PROPOSITION. _Let_ $c:A \longrightarrow A \otimes^* A$ _be a_ C*-_bigebra with identity. The following statements are equivalent:_

(i) _In_ bi[C^*_a] _the bigebra_ c _is a direct limit of finite dimensional_ C*-_bigebras_.

(ii) _The function_ λ:Spec A $\longrightarrow$ Spec E(A) _is bijective_ (i.e. _the idempotents of_ A _separate the points of_ Spec A).

(iii) Spec A _is a projective limit of finite semigroups_.

(iv) Spec A _is totally disconnected_.

Proof. (i) and (iii) are equivalent by duality and (ii) and (iv) are equivalent by duality. But, as is well known, (iii) and (iv) are equivalent ([11], p. 52).

REMARK. The maps in the inverse system in (iii) whose projective limit is Spec A, are all surjective. This means that the C*-morphisms of the direct system in (i), whose colimit is A, are injective. In this sense we may formulate (i) as

(i') The bigebra A is a union of finite dimensional subbigebras.

The condition, that the (central) idempotents of an algebra separate the maximal modular ideals is sometimes referred to as "weak biregularity."

Now we utilize the concept of primitive elements to introduce semicharacters.

Let $c: A \longrightarrow A \otimes^* A$ the dual C*-cogebra of a locally compact proper semigroup S. We may identify A with C(S) and $A \otimes^* A$ with $C(S \times S)$, so that $c = C(m)$ where $m: S \times S \longrightarrow S$ is the semigroup multiplication. Thus $f: S \longrightarrow \mathbb{C}$ is in P(C(S)) (see 10.18) if and only if $c(f)(s,t) = (f \otimes f)(s,t)$.

But $c(f)(s,t) = f(st)$ and $(f \otimes f)(s,t) = f(s)f(t)$. The primitive elements are thus exactly the topological semigroup morphisms $S \longrightarrow \mathbb{C}$. Since S is compact, so is f(S), hence f(S) is contained in the unit disc D, which also follows from $\|f\| \leq 1$, see 10.18. Thus, if $\mathcal{C}$

denotes the category of compact semigroups, $P(C(S)) = \mathcal{C}(S,D)$.

11.9. DEFINITION. The primitive elements of $C(S)$ are called *the semicharacters of* S.

11.10. If A is the dual C*-bigebra of a compact semigroup S, then $A_p = A$ iff the semicharacters of S separate points, where A_p = closure of the linear span of P(A).

We are now in a position to complete Theorem 10.18.

11.11. THEOREM. *Let* $c:A \longrightarrow A \otimes^* A$ *be a* C*-*cogebra with coidentity with a non-zero primitive element* a. *Then* $\|a\| = 1$.

Proof. Since $\|a\| = 1$ iff $\|a^*a\| = 1$ and $a^*a \in P(A) \setminus \{0\}$ we may assume that a is hermitean. Let B be the C*-subalgebra generated by a (and 1 if A has an identity). Since B is flat as an abelian C*-algebra (7.9 (3)), we may consider $B \otimes^* B$ as a C*-subalgebra of $A \otimes^* A$. Since $c(a) = a \otimes a$ [$c(1) = 1 \otimes 1$ if c is a cogebra with identity] and B is generated by c [and 1] the algebra B is mapped into the algebra generated by $a \otimes a$ [and $1 \otimes 1$] under c; the latter is clearly contained in $B \otimes^* B$, hence c induces a cogebra $c|B:B \longrightarrow B \otimes^* B$ [with identity if c has an identity]. If $e:A \longrightarrow \mathbb{C}$ is the coidentity of A, then $e|B:B \longrightarrow \mathbb{C}$ is a coidentity of B.

It is therefore no loss of generality to assume that A is an abelian C*-cogebra, and by 10.14 we may assume that A has an identity. We may further assume that the C*-algebra A is generated by the hermitean elements a and 1. Thus, by duality, $A = C(S)$ for a compact semigroup S with identity. By the discussion preceding 11.9, $a:S \longrightarrow D$ is a semicharacter, i.e. a morphism of semigroups. Since a is the generator of the algebra, it is a homeomorphism from S onto the spectrum of a. Thus $1 \in a(S)$, since $a \neq 0$, which means $\|a\| \geqq 1$. Since $a(S) \subseteqq D$, we also have $\|a\| \leqq 1$. Note that a(S) is real since a is hermitean; thus $a(S) \cong S$ is in fact a subsemigroup of $[-1,1]$.

<u>EXAMPLE</u>. Let S be the additive compact semigroup $[c,\infty] \subseteq R \cup \{+\infty\}$, $0 \leq c$. The semicharacters are the functions $f_r:S \longrightarrow D$, $f_r(s) = e^{-rs}$ (with $e^{-\infty} = 0$), $r \geq 0$ and the constant function with value zero. If $0 < c$, then the only primitive element with norm 1 is the constant semicharacter with value 1. This example shows further that P(A) is not generally discrete; for if $A = C(S)$, then P(A) has the components $\{1\}$ and $\{f_r : r > 0\} \cup \{0\}$, if $c > 0$, and the components $\{1\}$, $\{f_r : r > 0\}$, $\{0\}$ if $c = 0$.

We conclude the section by observing that the part of the duality theory referring to proper locally compact semigroups is a comparatively irrelevant generalization over the case of compact semigroups.

11.12. Let $c: A \longrightarrow A \otimes^* A$ be a C*-cogebra and S its dual proper locally compact semigroup. Let $\tilde{A} = \mathbb{C} \oplus A$ and $\tilde{c}: \tilde{A} \longrightarrow \tilde{A} \otimes^* \tilde{A}$ be the C*-bigebra obtained from c by adjoining an identity (10.14). Let $\tilde{S}$ be the dual semigroup of $\tilde{A}$. If S is not compact, then $\tilde{S}$ is the one-point compactification $S \cup \{\infty\}$ of S and the point ∞ acts as zero. If S is compact, then $\tilde{S}$ is obtained from S by adjoining an isolated zero.

Proof. The topological structure of $\tilde{S}$ is well known; also $\tilde{S}$ is a topological compact semigroup. It remains to show that ∞ acts as a zero. But $\{\infty\}$ is canonically identified with Spec $\tilde{A}/A$ according to 11.2, and is an ideal of $\tilde{S}$ by 11.2. iii). This is the assertion.

<u>From this point of view it is no restriction of generality to concentrate on compact semigroups and C*-bigebras with identity</u>.

One might formulate this observation more rigorously by saying that the functor $c \longmapsto \tilde{c}$ coreflects the category of commutative C*-cogebras into the category of commutative C*-bigebras with identity and that $S \longrightarrow \tilde{S}$ coreflects the category of locally compact proper topological semigroups into the category of compact topological semigroups.

We leave the following as an exercise:

11.13. Let $c: A \longrightarrow A \otimes A$ be a C*-cogebra and S its dual semigroup. Let $c^1: A^1 \longrightarrow A^1 \otimes A^1$ be the C*-cogebra obtained from c by adjoining a coidentity (10.15). Let

S^1 be the dual semigroup of A^1. Then S^1 is the topological semigroup obtained from S by adjoining a discrete identity.

12. Involutive bigebras and their relation to C*-bigebras

We recall that Inv is the category of involutive algebras over $\mathbb{C}$ and that the ordinary tensor product makes Inv into a multiplicative category. An Inv-cogebra is a cogebra over Inv, i.e. a morphism $c:A \longrightarrow A \otimes A$ of involutive algebras. The remarks made after the definition of C*-cogebras in 10.1 maintain in the present situation.

However, one might add that the distinction between cogebras and bigebras perhaps need not be as meticulously observed as in the case of C*-algebras. For if $c:A \longrightarrow A \otimes A$ is an Inv-cogebra, there is a morphism $m:A \otimes A \longrightarrow A$ of involutive vector spaces such that $m(a \otimes b) = ab$. The diagram $A \xrightarrow{c} A \otimes A \xrightarrow{m} A$ indeed satisfies the commutative diagram of 5.1, and an Inv-cogebra is automatically a bigebra over the category of involutive vector spaces. This was not the case with C*-cogebras.

We recall that for a C*-algebra A the C*-tensor-product contains the algebraic tensor product $A \otimes A$. It is, however, not to be expected and it is not true, in general, that for a C*-cogebra $c:A \longrightarrow A \otimes^* A$ the comultiplication factors through $A \otimes A$. However, in the following we construct functorially an involutive subcogebra $c|A':A' \longrightarrow A' \otimes A'$.

12.1. <u>THEOREM</u>. <u>Let</u> $c: A \longrightarrow A \otimes^* A$ <u>be a</u> C*-<u>cogebra and</u> A' <u>an involutive subalgebra of</u> A. <u>Construct in</u> Inv <u>by transfinite induction the following sequence</u> A_n <u>of subalgebras</u>: <u>Suppose that</u> n <u>is an ordinal and that</u> A_n <u>was defined for</u> $m < n$. <u>If</u> n <u>is a limit ordinal, let</u> $A_n = \cap\{A_m : m < n\}$. <u>If</u> $n = 0$, <u>let</u> $A_n = A$. <u>Otherwise, if</u> $n = m + 1$, <u>let</u> $A_n = A_m \cap c^{-1}(A_m \otimes A_m)$. <u>Let</u> N <u>be an ordinal such that</u> $A_{N+1} = A_N$. (<u>Such an ordinal always exists</u>.) <u>Then the following statements are equivalent</u>:

1) $A' = A_N$.

2) A' <u>is the unique largest involutive subalgebra of</u> A <u>such that</u> $c(A') \subset A' \otimes A'$.

Proof. As a first step, we show that an A' as described in 2) actually exists. We have the following two lemmas:

<u>LEMMA 1</u>. If V_i are involutive subvectorspaces of A satisfying $c(V_i) \subset V_i \otimes V_i$, $i \in I$ and if $V = \Sigma\{V_i : i \in I\}$, then $c(V) \subset V \otimes V$.

<u>LEMMA 2</u>. If V is an involutive subvector space of A with $c(V) \subset V \otimes V$ and A' the smallest involutive subalgebra containing V (namely the set of all finite linear combinations of finite products of elements of V) then $c(A') \subset A' \otimes A'$.

Lemma 1 is a consequence of the distributivity of $\otimes$, the second lemma follows from the fact that c is a morphism of algebras and the fact that the algebra generated by $V \otimes V$ in $A \otimes A$ is contained in $A' \otimes A'$.

The set of all involutive subvector spaces V with $c(V) \subset V \otimes V$ contains $\{0\}$ (and, if c is a C*-bigebra with identity, the subalgebra $\mathbb{C}.1$). It is clearly inductive. By Lemma 2 any maximal element is in fact a subalgebra. By Lemma 1 there is at most one maximal element. Thus there is a unique maximal involutive subalgebra A' with $c(A') \subset A' \otimes A'$ and if c has an identity, then A' contains the identity. Let now A' remain this algebra. Now we show $A_N = A'$. Since $A_{n+1} = A_n \cap c^{-1}(A_n \otimes A_n)$ for all n, it is clear that $A' \subset A_n$ for all n. Conversely, $c(A_N) = c(A_{N+1}) \subset A_N \otimes A_N$, so A_N is an involutive subalgebra satisfying $c(A_N) \subset A_N \otimes A_N$ and is, therefore, contained in A'. Thus $A_N = A'$.

NOTATION. We write $c': A' \longrightarrow A' \otimes A'$ for the restriction and corestriction of c.

12.2. THEOREM. *With the notation of the previous theorem* $A \longrightarrow A'$ *is a functor from the category of* C*-*cogebras into the category of* Inv-*cogebras*.

Proof. The proof is simple: If

$$\begin{array}{ccc} A & \xrightarrow{\;c\;} & A \otimes^* A \\ \varphi \downarrow & & \downarrow \varphi \otimes^* \varphi \\ B & \xrightarrow[\;d\;]{} & B \otimes^* B \end{array}$$

is a morphism of C*-bigebras, then $\varphi(A')$ is an involutive subalgebra of B such that $d(\varphi(A)) = (\varphi \otimes^* \varphi)c(A') = (\varphi \otimes^* \varphi)(A' \otimes A') = \varphi(A) \otimes \varphi(A)$. Thus $\varphi(A) \subset B'$ since B' is the unique largest involutive subalgebra with

$d(B') \subset B' \otimes B'$. If we define $\varphi':A' \longrightarrow B'$ to be the restriction and corestriction of φ then

$$\begin{array}{ccc} A' & \xrightarrow{\ c'\ } & A' \otimes A' \\ {\scriptstyle \varphi'}\downarrow & & \downarrow{\scriptstyle \varphi' \otimes \varphi'} \\ B' & \xrightarrow[\ d'\ ]{} & B' \otimes B' \end{array}$$

The remaining observations are straightforward.

12.3. DEFINITION. Let $c:A \longrightarrow A \otimes A$ be a C*-cogebra. We call c *saturated* if A' is dense in A.

12.4. THEOREM. *Let* $c:A \longrightarrow A \otimes^* A$ *be a C*-cogebra. Then* c *induces a C*-cogebra* $\overline{c'}:\overline{A'} \longrightarrow \overline{A'} \otimes^* \overline{A'}$ *which is saturated, and* $\overline{A''} = \overline{A'}$.

Proof. Clearly $\overline{A'}$ is a C*-subalgebra of A. Since $\overline{A'} \otimes^* \overline{A'}$ and $\overline{A' \otimes A'}$ both are C*-subalgebras of $A \otimes^* A$ containing the subalgebra $A' \otimes A'$ densely, they agree. Since $c(A') \subset A' \otimes A'$, then $c(\overline{A'}) \subset \overline{c(A')} \subset \overline{A' \otimes A'}$, the restriction and corestriction $\overline{c'}$ of c to $\overline{A'} \longrightarrow \overline{A'} \otimes^* \overline{A'}$ is indeed a C*-bigebra. Since A' is an involutive subalgebra of $\overline{A'}$ satisfying $\overline{c'}(A') = c(A') \subset A' \otimes A'$, we have $A' \subset \overline{A''}$. On the other hand, $\overline{A''}$ is an involutive subalgebra of A satisfying $c(\overline{A'}) = \overline{c'}(\overline{A''}) \subset \overline{A''} \otimes \overline{A''}$, therefore we have $\overline{A''} \subset A'$. Thus equality holds and $\overline{A'}$ is saturated.

The next question naturally is whether there is a route leading in the other direction: Given an involutive cogebra, is there functorially associated with it a

C*-bigebra? Perhaps one would like to say more than we are going to discuss, but for the time being we declare it satisfactory to understand at least the commutative situation.

Firstly, we have to duplicate some observations already made in the context of commutative C*-algebras. Let A be an involutive commutative algebra over the field of complex numbers. With Spec A we denote <u>the spectrum of</u> A, i.e. the set of all algebra characters, i.e. non-zero morphisms $f:A \longrightarrow \mathbb{C}$ of involutive algebras with the topology of point wise convergence, i.e. the topology induced on Spec $A \subset \mathbb{C}^A$ by the product topology.

It is now clear that Spec A is a completely regular topological space, and that for every $a \in A$ the function $\hat{a}:\text{Spec } A \longrightarrow \mathbb{C}$ defined by $\hat{a}(f) = f(a)$ is continuous. The map $a \longmapsto \hat{a}:A \longrightarrow C(\text{Spec } A)$ is clearly a morphism of involutive algebras into the algebra of complex valued continuous functions on Spec A. Its kernel is $\cap\{\ker f:f \in \text{Spec } A\}$, and we call it <u>the Gelfand-representation</u>.

We note in passing that Spec A may be identified with the set of all maximal modular involutive ideals I of A for which the quotient algebra A/I is isomorphic to $\mathbb{C}$. As such, it has another natural topology, namely the hull kernel topology; the two topologies are, in general, different.

A morphism $\varphi:A \longrightarrow B$ of involutive commutative algebras will be called <u>proper</u> if for every $f \in \text{Spec } B$ the composition $f\varphi$ is not zero (i.e. if no maximal ideal of A corresponding to an element of Spec A is every mapped <u>onto</u> B under φ). Note that this condition is automatically satisfied if A and B have identities which are respected by φ. If $\varphi:A \longrightarrow B$ is a proper morphism, then there is a function $\text{Spec } \varphi:\text{Spec } B \longrightarrow \text{Spec } A$ such that $(\text{Spec } \varphi)(f) = f\varphi$ for $f \in \text{Spec } B$. By the definition of the topologies on the spectra, Spec φ is a continuous function. Thus Spec is a cofunctor for the category of involutive commutative algebras with proper morphisms into the category of completely regular spaces.

12.5. PROPOSITION. *Let A_i, $i = 1, 2$ be commutative involutive algebras. Then the function $\psi:\text{Spec } A_1 \times \text{Spec } A_2 \longrightarrow \text{Spec } A_1 \otimes A_2$ defined by $\psi(f_1, f_2) = f_1 \otimes f_2$ is a continuous bijection. If for each $f_i \in \text{Spec } A_i$ there is some element $e_i \in A_i$ with bounded Gelfand transform and $f_i(e_i) \neq 0$, $i = 1, 2$, then ψ is a homeomorphism. (This applies particularly to the case that A_i has an identity.)*

REMARK. *It is in this sense that Spec is a multiplicative functor.*

Proof. For any pair A_i, $i = 1, 2$ of complex vector spaces there is a canonical injection $\text{Vect}(A_1, \mathbb{C}) \otimes \text{Vect}(A_2, \mathbb{C}) \longrightarrow \text{Vect}(A_1 \otimes A_2, \mathbb{C})$ under which the element $f_1 \otimes f_2$, $f_i: A_i \longrightarrow \mathbb{C}$ is identified with a functional $A_1 \otimes A_2 \longrightarrow \mathbb{C}$

given by $(f_1 \otimes f_2)(a_1 \otimes a_2) = f_1(a_1)f_2(a_2)$. If A_i, $i = 1, 2$ are algebras and the f_i, $i = 1, 2$ are multiplicative, then clearly $f_1 \otimes f_2$ is multiplicative. Conversely, if $F: A_1 \otimes A_2 \longrightarrow \mathbb{C}$ is multiplicative, then there are $e_i \in A_i$ such that $F(e_1 \otimes e_2) = 1$; define $f_1(a_1) = F(a_1e_1 \otimes e_2)$ and $f_2(a_2) = F(e_1 \otimes e_2a_2)$. Then $f_1(a_1)f_1(b_1) = F(a_1e_1 \otimes e_2)F(b_1e_1 \otimes e_2) = F(a_1b_1e_1{}^2 \otimes e_2{}^2) = F(a_1b_1e_1 \otimes e_2)F(e_1 \otimes e_2) = f_1(a_1b_1)1$. Thus f_1 is multiplicative, and so is f_2. Moreover, $f_1(a_1)f_2(a_2) = F(a_1e_1 \otimes e_2a_2) = F(a_1 \otimes a_2)F(e_1 \otimes e_2) = F(a_1 \otimes a_2)$. Thus $F = f_1 \otimes f_2$, and we conclude that the multiplicative functionals of $A_1 \otimes A_2$ are exactly the ones of the form $f_1 \otimes f_2$ with $f_i \in \text{Spec } A_i$.

Thus ψ is surjective. Now suppose that $f_1 \otimes f_2 = g_1 \otimes g_2$; then $f_1(a_1)f_2(a_2) = g_1(a_1)g_2(a_2)$ for all $a_m \in A_m$, $m = 1, 2$. Now choose a_2 so that $g_2(a_2) = 1$ and let $c = f_2(a_2)$. Then f_1 and $g_1 = cf_1$ are multiplicative functionals. In particular $g_1(a^n) = cf_1(a^n)$, $g_1(a^n) = g_1(a)^n = c^nf_i(a)^n = c^nf_i(a^n)$. This implies $c^n = 1$ for $n = 1,2,3,\ldots$. Thus $c = 1$ and $f_1 = g_1$. Similarly $g_2 = f_2$.

Finally we discuss the continuity of ψ. Let $(f_i,g_i)_{i \in I}$ be an arbitrary net and (f,g) an arbitrary element on $\text{Spec } A_1 \times \text{Spec } A_2$. Consider the statements

(a) $\lim(f_i,g_i) = (f,g)$

(a') $\lim f_i(a) = f(a)$ and $\lim g_i(b) = g(b)$ for all $a \in A_1$, $b \in A_2$,

(b) $\lim \psi(f_i, g_i) = \psi(f, g)$

(b') $\lim f_i(a)g_i(b) = f(a)g(b)$ for all $a \in A_1$, $b \in A_2$.

Then (a) <==> (a') and (b) <==> (b'). Also (a') ==> (b') is trivial.

Now suppose that $e_m \in A_m$ has a bounded Gelfand transform, $m = 1, 2$, with $f(e_1) \neq 0$, $g(e_2) \neq 0$. Then there is a subnet $(f_{i(j)}, g_{i(j)})$ such that $(c,d) = \lim f_{i(j)}(e_1), g_{i(j)}(e_2))$ exists. It follows then from (b') that $(\lim f_{i(j)}(a))d = f(a)g(e_2)^*$ for all $a \in A_1$. Since $f(e_1) \neq 0$, then d cannot be 0; now $a \longmapsto d^{-1}g(e_2)f(a)$ is a limit of multiplicative functionals and is, therefore, a multiplicative functional, which implies $d^{-1}g(e_2) = 1$. This and a parallel argument show $c = f(e_1)$, $d = g(e_2)$, and $\lim f_{i(j)}(a) = f(a)$, $\lim g_{i(j)}(b) = g(b)$ for all $a \in A_1$, $b \in A_2$. But now it turns out that the limits are completely independent of the subnet we chose. Hence condition (a') follows.

This immediately implies the following:

12.6. <u>THEOREM</u>. <u>Let</u> $c: A \longrightarrow A \otimes A$ <u>be a commutative proper</u> Inv-<u>cogebra</u> (i.e. c <u>is a proper morphism</u>). <u>Then</u> Spec A <u>is a completely regular topological semigroup relative to a multiplication</u> m <u>which makes the following diagram commutative</u>.

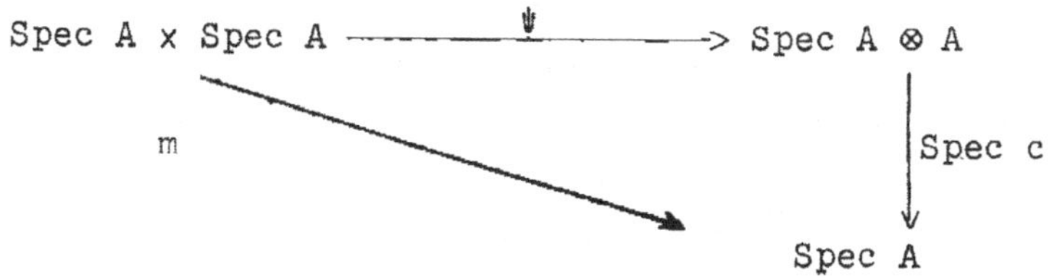

REMARK. *If c is a cogebra with identity, then c is automatically proper.*

We thus have

12.7. COROLLARY. *Spec:co[$\mathrm{Inv}_{\mathrm{prop}}$] $\longrightarrow \mathcal{CR}$ is a cofunctor from the category $\mathrm{Inv}_{\mathrm{prop}}$ of commutative proper Inv-cogebras and proper morphisms into the category $\mathcal{CR}$ of completely regular semigroups.*

The next step is to identify some of the functors introduced earlier in the section in different terms.

Let S be a locally compact proper semigroup. For each $f \in C_o(S)$, since S is proper, the translates $x \longmapsto f(sx)$, $f(xs)$, $f(sxt)$ belong to $C_o(S)$. Let $\langle f\rangle_\ell$ (respectively, $\langle f\rangle_r$, respectively, $\langle f\rangle$) denote the subvector space spanned by the translates $x \longmapsto f(xs)$ (respectively, $x \longmapsto f(sx)$, respectively $x \longmapsto f(sxt)$) in $C_o(S)$. If $c:C_o(S) \longrightarrow C_o(S) \otimes^* C_o(S) = C_o(S \times S)$ is the corresponding comultiplication, then $g \in \langle f\rangle_\ell$ implies $g(x) = f(xs)$ for some $s \in S$. Then $c(g)(x, y) = g(xy) = f(xys)$. Suppose that $\{f_i : i \in I\}$ is a basis of $\langle f\rangle_\ell$. Then since $x \longmapsto c(g)(x, y)$ is in $\langle f\rangle_\ell$, we have $c(g)(x, y) = \Sigma\ \{h_i(y)f_i(x) : i \in I\}$ for suitable functions $h_i : S \longrightarrow \mathbb{C}$, the sum being in reality a finite one.

We recall the following elementary lemma:

LEMMA. If F is a vector space of functions $\varphi : S \longrightarrow K$ into a field K, then there is a subset S_F of S such that $\varphi \longmapsto \varphi|S_F$ is an isomorphism from F onto K^{S_F}.

In the light of this lemma, we may actually assume that $I \subset S$ and that the basis elements have the property that $f_i(j) = 0$ if $i \neq j$ and $f_i(i) = 1$ for $i, j \in I$. Then $f(iys) = c(g)(i, y) = h_i(y)$ which shows that all h_i are in fact in $\langle f^i \rangle$ where $f^i(x) = f(ix)$. Since the function $(x, y) \longmapsto f_i(x)h_i(y)$ is identified with $f_i \otimes h_i$, we have found that $c(g) = \Sigma \{f_i \otimes h_i : i \in I\}$, where we have in fact a finite sum as soon as it is evaluated at a particular element $(x, y) \in S \times S$.

Now suppose that I is finite, so that the sum is finite as it stands. This is the case iff $\dim \langle f \rangle_\ell < \infty$, and this in turn implies $\dim \langle f^s \rangle_\ell < \infty$ for all $s \in S$, where $f^s(x) = f(sx)$.

Let $R_\ell(S) = \{f \in C_o(S) : \dim \langle f \rangle_\ell < \infty\}$. Then $c(g) \in R_\ell(S) \otimes R_\ell(S)$ for all $g \in \langle f \rangle_\ell$. Thus $c(R_\ell(S)) \subseteq R_\ell(S) \otimes R_\ell(S)$.

Clearly, $R_\ell(S)$ is invariant under the involution. The function $(f, g) \longmapsto fg$ is a bilinear map. Thus $\langle fg \rangle_\ell$ is a homomorphic image of $\langle f \rangle_\ell \otimes \langle g \rangle_\ell$ and is, therefore, finite dimensional, provided $f, g \in R_\ell(S)$.

Thus, if we denote with c_ℓ the appropriate restriction and corestriction of c, we know that $c_\ell : R_\ell(S) \longrightarrow R_\ell(S) \otimes R_\ell(S)$ is an Inv-cogebra and a sub-cogebra of $C_o(S)$. This implies $R_\ell(S) \subseteq C_o(S)'$ by 12.1.

If $R_r(S) = \{f \in C_o(S)$: the vector space spanned by the translates $x \longmapsto f(sx)$ is finite dimensional$\}$, then $R_r(S)$ consists exactly of the coefficients of the finite dimensional antirepresentations and $c_r : R_r(S) \longrightarrow R_r(S) \otimes R_r(S)$ again is an Inv-cogebra in $C_o(S)'$.

If $R_t(S) = \{f \in C_o(S)$: the vector space spanned by all translates $x \longmapsto f(sxt)$ is finite dimensional$\}$, then one still observes that $c_t : R_t(S) \longrightarrow R_t(S) \otimes R_t(S)$ is a sub-Inv-cogebra of $C_o(S)'$.

Now assume that $f \in C_o(S)'$. Then $c(f) \in C_o(S) \otimes C_o(S)$ so that $c(f) = \Sigma \{f_i \otimes h_i : i \in I\}$ with some finite set I, and f_i, $h_i \in C_o(S)$. For $s \in S$ we obtain $f(sx) = c(f)(s,x) = \Sigma \{f_i(s)h_i(x) : i \in I\}$. Thus the space of translates $x \longmapsto f(sx)$ is spanned by the finitely many elements h_i. Hence $f \in R_r(S)$. Hence $C_o(S)' \subseteq R_r(S) \cap R_\ell(S) \cap R_t(S) \subseteq R_r(S) + R_\ell(S) + R_t(S) \subseteq C_o(S)'$. Thus $C_o(S) = R_r(S) = R_\ell(S) = R_t(S)$.

If V is a finite dimensional vector space over C, and $\pi : S \longrightarrow \mathrm{End}(V)$ a continuous representation vanishing at ∞, then for any vector $v \in V$ and any linear form $w \in \hat{V}$ on V the function $s \longmapsto \langle\pi(s)v, w\rangle$ is seen to be in $R_\ell(S)$.

Conversely, suppose that $f \in R_\ell(S)$. Let $V = \langle f\rangle_\ell$ and define $\pi : S \longrightarrow \mathrm{End}\ V$ by $(\pi(s)g)(x) = g(xs)$. Then π is a continuous representation of S on V vanishing at ∞. _Now assume that_ S _has a left identity_ e. Let $w \in \hat{V}$ be defined by $\langle g,w\rangle = g(e)$. Then $\langle\pi(s)f,w\rangle = f(es) = f(s)$.

Thus we have the following theorem:

12.8. _THEOREM_. _Let_ S _be a locally compact proper semigroup_. _Let_ $f \in C_o(S)$. _Then the following statements are equivalent_:

(1) $f \in C_o(S)'$ (*see* 12.1).

(2) *The vector space spanned by the translates* $x \longmapsto f(xs)$ *is finite dimensional.*

(3) *The vector space spanned by the translates* $x \longmapsto f(sx)$ *is finite dimensional.*

(4) *The vector space spanned by the translates* $x \longmapsto f(sxt)$ *is finite dimensional.*

Any of these conditions implies any of the following:

(5) [(6)] *There is a finite dimensional continuous representation* [*antirepresentation*] $\pi : S \longrightarrow \mathrm{End}\ V$ *vanishing at* ∞ *so that* $f(s) = \langle \pi(s)\ v, w \rangle$ *for suitable* $v \in V$, $w \in \hat{V}$ *and all* $s \in S$.

If S *has an identity, then* (5) *or* (6) *implies* (1).

REMARK. If S is compact, then the condition "vanishing at ∞" is void.

12.9. DEFINITION. The involutive algebra $C_o(S)'$ is called R(S), *the representation* (*or Grothendieck*) *algebra of* S. The commutative Inv-cogebra $c' : R(S) \longrightarrow R(S) \otimes R(S)$ is called *the representation cogebra* (*or bigebra*) *of* S.

If S is compact, then $C_o(S) = C(S)$ has an identity, in which case $R(G) = C(S)'$ contains at least $\mathbb{C}$.1. Let $S = [0,\infty[$ with the operation $(x,y) \longrightarrow \max\{x,y\}$. Then S is a proper locally compact semilattice without any non-constant finite dimensional linear representations. Hence $R(S) = \{0\}$. If (S,b) is an arbitrary compact space with base point b and with multiplication $xy = b$ for all $x,y \in S$, then $R(S) = C(S)$.

If S is an arbitrary connected left zero semigroup (i.e. an arbitrary connected compact space with $xy = x$, $x,y \in S$), then $R(S) = C(S)$.

12.10. DEFINITION. A linear representation π of a locally compact semigroup is called *tapered*, if it satisfies the following conditions:

a) π is continuous.

b) π vanishes at infinity, if S is not compact.

12.11. PROPOSITION. *If S is a locally compact proper topological semigroup, then the C*-cogebra $C_o(S)$ is saturated in the sense of 12.3 if the finite dimensional tapered linear representations of S separate the points of S. If S has a right or left identity, or if $xs = ys$ and $sx = sy$ for all s implies $x = y$, then the converse holds.*

Proof. Clearly $R(S)$ is an involutive subalgebra of $C_o(S)$ containing the constants in the case that S is compact. By the theorem of Stone and Weierstrass $R(S)$ is dense iff it separates the points of S. If $\pi : S \longrightarrow \mathrm{End}(\mathbb{C}^n)$ is a tapered representation, then for any $v \in \mathbb{C}^n$ and any linear form w the function $s \longrightarrow \langle \pi(s)v,w \rangle$ is in $R(S)$. Thus, if the finite dimensional tapered representations separate the points, then so does $R(S)$. Conversely, if $R(S)$ separates, then the family $\{\langle f \rangle : f \in R(S)\}$ is a set of finite dimensional tapered left S-modules relative to the module operations $(s.g)(x) = g(xs)$ respectively $(g.s)(x) = g(sx)$.

If $x,y \in S$ are not separated by this family of modules, then for all $f \in R(S)$ we have $f(xs) = f(ys)$, $f(sx) = f(sy)$ for all $s \in S$. Since $R(S)$ separates points, this means $xs = ys$ and $sx = sy$ for all $s \in S$. Now any one of the two conditions is sufficient to ensure $x = y$:

(a) S has a left--or right--identity.

(b) $sx = sy$ and $xs = ys$ for all s implies $x = y$.

12.12. DEFINITION. a) A topological locally compact proper semigroup is called *a Peter-Weyl semigroup* if $R(S)$ is dense in $C_o(S)$, i.e. if $C_o(S)$ is a saturated C*-bigebra.

b) An arbitrary topological semigroup is called *representable*, if its finite dimensional continuous representations separate its points.

By 12.10 any locally compact proper Peter-Weyl semigroup with left or right identities is representable. A compact semigroup with left or right identities is a Peter-Weyl semigroup if and only if it is representable.

Every compact group, of course, is a Peter-Weyl semigroup because of the theorem of Peter and Weyl which says exactly that $R(S)$ separates the points. The semilattice on the unit interval (with the min operation) is a compact commutative semigroup with identity in which every finite dimensional linear representation is constant and $R(S)$ contains only the constants.

12.13. THEOREM. *Let* S *be a proper locally compact semigroup. Then there is a quotient semigroup* $W(S)$ *with a proper quotient morphism* $\pi: S \longrightarrow W(S)$ *such that*

i) *$W(S)$ is a Peter-Weyl semigroup.*

ii) *If T is a Peter-Weyl semigroup and $f:S \longrightarrow T$ a proper morphism, then there is a unique morphism $f':W(S) \longrightarrow T$ such that $f = f'\pi$.*

REMARK. For compact semigroups the full subcategory of compact Peter-Weyl semigroups is a complete subcategory, and the coadjoint functor theorem allows us to make the conclusion of the preceding theorem directly.

Proof. Let $c:A \longrightarrow A \otimes^* A$ be the dual of S. Let $W(S)$ be the dual of $\overline{c'}:\overline{A'} \longrightarrow \overline{A'} \otimes^* \overline{A'}$. Since $A' = R(\dot{S}) \subset C_o(S) = A$ and $\overline{A'} = \overline{R(S)}$ we obtain $W(S)$ as a quotient of S under a proper morphism by duality. Now suppose that $f:S \longrightarrow T$ is a proper morphism into a Peter-Weyl semigroup. Let $b:B \longrightarrow B \otimes^* B$ be the dual of T and

$$\begin{array}{ccc} B & \longrightarrow & B \otimes^* B \\ \downarrow \varphi & & \downarrow \varphi \otimes^* \varphi \\ A & \longrightarrow & A \otimes^* A \end{array}$$

the morphism dual fo f. Since ' is a functor, we obtain a diagram

$$\begin{array}{ccc} B' & \xrightarrow{b'} & B' \otimes B' \\ \varphi' \downarrow & & \downarrow \varphi' \otimes \varphi' \\ A' & \xrightarrow[c']{} & A' \otimes A' \end{array}$$

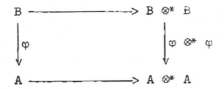

It is clear that there is no loss in assuming that f has dense image. This implies that φ and therefore φ' are injective. We consider every algebra in sight as sub-algebra of $A = C_o(S)$ and form closures. Since T is a Peter-Weyl semigroup, b is a saturated C*-bigebra. Thus $\overline{B'} = B$. Hence we have

$$\begin{array}{ccc}
B & \longrightarrow & B \otimes^* B \\
\downarrow & & \downarrow \\
\overline{A'} & \longrightarrow & \overline{A'} \otimes^* \overline{A'} \\
\downarrow & & \downarrow \\
A & \longrightarrow & A \otimes^* A
\end{array}$$

We let f' be the dual of $B \longrightarrow \overline{A'}$. Assertion ii) then follows by duality.

The preceding theorem explains via duality the significance of the functor $c \longrightarrow \overline{c'}$ of commutative C*-bigebras. How much can we say about the functor which associates with $c':A' \longrightarrow A' \otimes A'$ the completely regular semigroup Spec A', if anything? Notably, if $A' = R(S)$, how are S and Spec R(S) related?

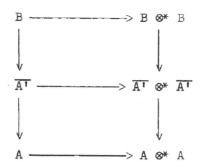

Take a commutative C*-bigebra $c:A \longrightarrow A \otimes^* A$. In view of the preceding theorem we may assume that it is saturated, and in view of 11.12 and the following remarks we may just as well assume that c has an identity. Let S be its dual semigroup. The spectrum Spec A' of the derived Inv-cogebra $c':A' \longrightarrow A' \otimes A'$ is a completely regular topological semigroup S' by 12.6 which contains (an isomorphic copy of) $S = \text{Spec } A = \text{Spec } \overline{A'}$, since A' separates the points of S. The functions $f \in R(S)$ extend uniquely to functions on S'; in other words A' is isomorphic to an involutive subalgebra $\tilde{R}(S')$ of the algebra $\tilde{C}(S')$ of all continuous functions on S' such that $\tilde{R}(S')|S = R(S)$. It remains true that for $F \in \tilde{R}(S')$ and $s',t' \in S'$ we have $F(s't') = (s' \otimes t')c'(F) = (s' \otimes t')\Sigma(F_i \otimes G_i) = \Sigma F_i(s')G_i(t')$ with suitable functions $F_i, G_i \in \tilde{R}(S)$. The arguments in the proof of 12.8 will then show that $\dim \langle F\rangle < \infty$ where $\langle F\rangle$ is the vector subspace of $\tilde{R}(S')$ generated by the translates of F. Let $\pi: S \longrightarrow \text{End}(V)$ be a finite dimensional continuous representation of S on a vector-space V. As v ranges through V and w through the dual $\hat{V}$ the representation is completely determined by the functions $s \longmapsto \langle\pi(s)v,w\rangle$. all of which are in R(S) hence extend uniquely to functions on S'. Define $\pi': S' \longrightarrow V^V$ so that $s \longmapsto \langle\pi'(s)v,w\rangle$ is the unique extension to a function on S'. It is readily checked that π' is a representation, and since all of these coefficient functions are continuous, π' is in fact a continuous representation $S' \longrightarrow \text{End } V$ extending π.

Conversely, assume that T is a completely regular topological semigroup containing S such that the following conditions are satisfied:

(i) The finite dimensional continuous representations of T separate points.

(ii) Each finite dimensional representation of S extends uniquely to a continuous representation of T.

(iii) S has an identity which is also the identity of T.

Let $\tilde{R}(T) = \{f$: there is a continuous representation $\pi:T \longrightarrow \text{End } V$ over a finite dimensional vector space V, an element $v \in V$ and a linear form w such that $f(t) = \langle\pi(t)v,w\rangle\}$. By (ii) the restriction $f \longmapsto f|S$ is an isomorphism $\tilde{R}(T) \longrightarrow R(S)$ of Inv-bigebras.

For each $t \in T$, the function $f \longmapsto f(t)$ is an element of Spec $\tilde{R}(T)$ and thus may be considered as an element of Spec $R(S)$. Thus there is an injection $i:T \longrightarrow S'$ such that $(f|S)^{\hat{}}(i(t)) = f(t)$ where $f \longmapsto \hat{f}$: $R(S) \longrightarrow \tilde{R}(S')$ is the Gelfand representation. Since S' has the topology of pointwise convergence, i is continuous. If $\tilde{\pi}$ is the unique extension of $\pi:S \longrightarrow \text{End } V$ to a representation of S', we also have $\langle(\pi|S)^{\sim}(i(t))v,w\rangle = \langle\pi(t)v,w\rangle$ for any continuous finite dimensional representation π of T, whence $(\pi|S)^{\sim}(i(t)) = \pi(t)$. This shows that i is a semigroup morphism.

If S is commutative, then C(S) and so R(S) is cocommutative, hence S' is commutative. If S is idempotent, then the C*-bigebra C(S) and so the Inv-bigebra R(S) is idempotent. Hence S' is idempotent. (Later on we will show that if S is a group [semilattice] then S' is a group [semilattice]; in fact it will turn out that S' = S in this instance).

Suppose that $S = \varprojlim S_i$, specifically that S is a projective limit of compact semigroups S_i. Then the C*-bigebra C(S) is an injective limit of the $C(S_i)$. However, as we shall see, R(S) need not be the limit of the $R(S_i)$.

We have proved the following theorem:

12.14. THEOREM. *There is a functor* $S \longrightarrow S'$ *from the category of compact Peter-Weyl semigroups and proper morphisms into the category of representable semigroups* (12.11) *such that*:

(1) $S \subseteq S'$ (*up to equivalence*).

(2) *If* T *is any subsemigroup of* S' *containing* S *then every continuous finite dimensional representation of* S *extends uniquely to a continuous representation of* S'.

(3) *If* T *is any representable semigroup with identity* 1 *containing* S *so that* $1 \in S$ *and that every finite dimensional representation of* S *extends uniquely to* T, *then there is a continuous injection* $i:T \longrightarrow S'$ *fixing* S *elementwise*.

(4) *If* A *is the dual* C*-*bigebra of* S, *then* $S' =$ Spec A'.

(5)(a) S *is commutative iff* S' *is commutative*.

(b) S *is idempotent iff* S' *is idempotent*.

(c) S *is a semilattice iff* S' *is a semilattice*.

The following observations complement 12.14.

12.15. PROPOSITION. *Let* $\alpha:S_1 \longrightarrow S_2$ *be an injective morphism of representable semigroups, such that all* continuous finite dimensional representations of S_1 factor uniquely through S_2. Then α is an epic in the category of representable semigroups.

Proof. Let $f,g:S_2 \longrightarrow T$ be morphisms of representable semigroups such that $f\alpha = g\alpha$. Let π be any finite dimensional continuous representation of T. Then $\pi f\alpha = \pi g\alpha$. By the uniqueness hypothesis this implies $\pi f = \pi g$. Since T is representable, $f = g$ follows.

12.16. COROLLARY. Let S be a compact Peter-Weyl semigroup with identity 1 and $S \subseteq T \subseteq S'$ as in 12.14. Then the inclusion $S \longrightarrow T$ is an epic in the category of representable semigroups.

12.17. PROPOSITION. Let S be a compact Peter-Weyl semigroup with identity 1 and T a representable semigroup such that $S \subseteq T$ and 1 is an identity of T such that every finite dimensional continuous representation of S extends uniquely to T. If $\pi:T \longrightarrow \text{End } V$ is a finite dimensional continuous representation of T, then the inclusion $\pi(S) \longrightarrow \pi(T)$ is an epic in the category of Peter-Weyl semigroups.

Proof. By 12.14 we may assume $S \subseteq T \subseteq S'$. Let G be a Peter-Weyl semigroup and $\alpha, \beta:\pi(T) \longrightarrow G$ two morphisms such that $\alpha|\pi(S) = \beta|\pi(S)$. The $\alpha\pi|S = \beta\pi|S$, and since $S \longrightarrow T$ is an epic by 12.16, we have $\alpha\pi = \beta\pi$. This, of course, implies $\alpha = \beta$.

Let us look at some simple examples. Let S be the semigroup $N \cup \{\infty\}$ under addition, where $N = \{0,1,\ldots\}$ is the semigroup of natural numbers under addition and $N \cup \{\infty\}$ its one point compactification with $s + \infty = \infty + s = \infty$ for $s \in S$.

If $\pi: N \longrightarrow \text{End } V$ is a finite dimensional representation of the semigroup of natural numbers, and if $\mathbb{C}[X]$ is the polynomial ring in one variable, then V is a $\mathbb{C}[X]$-module via $(\sum_{i=0}^{n} a_i X^i) \cdot v = \sum_{i=0}^{n} a_i \pi(i)(v)$. It is no loss of generality to assume $\pi(0) = 1$. Thus every representation of S defines a $\mathbb{C}[X]$-module (the converse holds iff $\lim X^i \cdot v = 0$ for all v and $i \longrightarrow \infty$). As is well known from the theory of the Jordan normal form of vector space endomorphisms, each finite dimensional $\mathbb{C}[X]$-module splits into a direct sum of modules; if V is one of them, then $v \longrightarrow X \cdot v$ has the form $\lambda\epsilon_n + \nu_n$, where ϵ_n is the identity morphism of the n-dimensional vector space V, where $\lambda \in \mathbb{C}$, and where ν_n is nilpotent and has, relative to a suitable basis $e_1,\ldots,e_n$ of V, the property that $\nu_n e_p = e_{p+1}$, where we define $e_g = 0$ for $g > n$. We conclude that every continuous representation of S decomposes into a direct sum of representations π each of which is of the form $\pi(s) = (\lambda\epsilon_n + \nu_n)^s$, $s = 0,1,\ldots$ for suitable $n = 1,2,\ldots,\lambda \in \mathbb{C}$, $|\lambda| < 1$, $\pi(\infty) = 0$. (The condition $|\lambda| < 1$ follows from continuity.) For natural s we have $(\lambda\epsilon_n + \nu_n)^s = \sum_{r=0}^{s} \binom{s}{r} \lambda^r \nu_n^{s-r}$. The coefficient functions of π are spanned by $s \longmapsto \langle \pi(s)e_i, \hat{e}_k \rangle$ with a dual basis

$\hat{e}_k$ to the basis e_i. We then have $<\pi(s)e_i, \hat{e}_k> = \sum_{r=0}^{s} \binom{s}{r}\lambda^r$ $<\nu_n^{s-r} e_i, \hat{e}_k> = \sum_{i=0}^{s} \binom{s}{r}\lambda^r <e_{i+s-r}, \hat{e}_k> =$

$$= \begin{cases} \binom{s}{s-(k-i)}\lambda^{s-(k-i)} & \text{for } 0 \leq k - i \leq s \\ 0 & \text{otherwise} \end{cases},$$

if $\lambda \neq 0$, and $<\pi(s)e_i,\hat{e}_k> = <\nu_n^{s} e_i, \hat{e}_k> = <e_{i+s}, \hat{e}_k> =$ $= \begin{cases} 1 & \text{if } s = k - i \\ 0 & \text{otherwise} \end{cases}$. Thus the coefficient space of this representation is spanned by the constants and the functions $f_{m,\lambda}$, $m = 0, \ldots, n$, $|\lambda| < 1$ given by

$$f_{m,\lambda}(s) = \begin{cases} \binom{s}{m}\lambda^{s-m} & \text{for } m \leq s \\ 0 & \text{otherwise} \end{cases}, \text{ if } \lambda \neq 0$$

$$f_{m,o}(s) = \begin{cases} 1 & \text{for } m = s \\ 0 & \text{otherwise} \end{cases} = \delta_{ms}.$$

It follows that $R(S)$ is generated as a ring by the functions $f_{m,\lambda}$, $m = 0, 1, 2, \ldots; |\lambda| < 1$. If I_m denotes the ideal $\{s \in S : s \geqq m\}$ of S, then $S = \varprojlim S/I_m$, $m = 0, 1, 2, \ldots$; now $R(S/I_m)$, when considered as a subalgebra of $R(S)$, consists of all functions which are constant on I_m. The direct limit of the $R(S/I_m)$ then is the subalgebra of all functions which are eventually constant. It does not contain any of the functions $f_{m,\lambda}$.

Now let $\pi_o : S' \longrightarrow \mathbb{C}^\times \cong \text{End}\, \mathbb{C}$ be the morphism of S' into the multiplicative semigroup of complexes given by $\pi_o(s) = e^{-s}$ for $s \in S$, $\pi_o(\infty) = 0$. Let $H^\times$ be the multiplicative semigroup of non-negative reals and let $\varphi : \mathbb{C}^\times \longrightarrow H^\times$ denote the morphism $\varphi(c) = |c|$. Let $R^* =$

$R \cup \{+\infty\}$ be the semigroup of additive reals to whose positive end the semigroup zero $+\infty$ is attached and let $\psi: H^{\times} \longrightarrow R^*$ be the isomorphism given by $\psi(h) = -\log h$, $\psi(0) = +\infty$. Then $\psi\varphi\pi_o: S' \longrightarrow R^*$ is the identity on S.

We define $\rho_{1,2}: R^* \longrightarrow \mathbb{C}^{\times}$ by $\rho_1(r) = e^{-r}$, $\rho_1(+\infty) = 0$ and by $\rho_2(r) = e^{-r+2\pi i r}$, $\rho_2(+\infty) = 0$. Then $\rho_{1,2}\ \psi\varphi\pi_o$: $S' \longrightarrow \mathbb{C}^{\times}$ are two representations agreeing on S. Hence, by 12.14, they agree. However, this implies that $\psi\varphi\pi_o(S') \subseteqq \{r \in R^*: \rho_1(r) = \rho_2(r)\} = S$. The identity morphism $S' \longrightarrow S'$ and the morphism $S' \xrightarrow{\psi\varphi\pi_o} S \xrightarrow{c} S'$ agree on S, hence they coincide by 12.14. This implies $S' = S$.

In many respects, the following example is simpler.

Let $S = H^*$ be the one point compactification semigroup of the additive half line $H = \{r \in R: r \geq 0\}$. Then any representation $\pi: H^* \longrightarrow \text{End } V$, $\pi(0) = 1$ defines a one-parameter subsemigroup $\pi|H: H \longrightarrow G\!\mid\!(V)$ into the Lie-group of all vector space automorphisms of V, and every such is of the form $\pi(h) = \exp h\alpha$ with a vector space endomorphism α of V. We must have $\lim \exp h\alpha = 0$ for $h \longrightarrow 0$. If we denote the semigroup $-H \cup H^*$ again with R^*, then clearly every representation extends uniquely to R^*. Thus $S \subseteq R^* \subseteq S'$. The representation $\pi_o: S \longrightarrow \mathbb{C}^{\times}$ ($\mathbb{C}^{\times} \cong \text{End } \mathbb{C}$ is the multiplicative semigroup of complex numbers) given by $\pi_o(s) = e^{-s}$, of course extends to R^* but also to S'. Let $H^{\times}$ be the multiplicative semigroup on the half line. Then $z \longrightarrow |z|: \mathbb{C}^{\times} \to H^{\times}$

is a morphism φ.

Let $\psi: H^X \longrightarrow R^*$ be the isomorphism defined by $\psi(h) = -\log h$ for $h \neq 0$ and $\psi(\infty) = \infty$; then $\psi\varphi\pi_o: S' \longrightarrow R^*$ is a morphism, whose restriction to R^* is the identity. In other words, R^* is a retract of S'. But now the identity $S' \longrightarrow S'$ and $S' \xrightarrow{\psi\varphi\pi_o} R^* \xrightarrow{c} S'$ agree on S. Hence they agree by 12.16. Thus $S' = R^*$.

If $\pi: R \longrightarrow \text{End } V$ is a one parameter group, then it is of the form $\pi(r) = \exp r\alpha$ with an endomorphism of the vector space V. By the theory of the Jordan normal form, $\alpha = \alpha_1 \oplus \ldots \oplus \alpha_m$ with $\alpha_k = \lambda_k \epsilon_{n(k)} + \nu_{n(k)}$, $k = 1, \ldots, m$, $\epsilon_{n(k)}$ = identity of the $n(k)$ - dimensional complex vector space, $\nu_{n(k)}$ nilpotent of the form described in the previous example. Thus

$$\pi(r) = e^{r\lambda_1} \exp r\, \nu_{n(1)} \quad \ldots \quad e^{r\lambda_m} \exp r\, \nu_{n(m)}.$$

If now $\lim \pi(r) = 0$ for $r \longrightarrow \infty$, we deduce

(a) $\text{Re } \lambda_k < 1$ for $k = 1, \ldots, m$.

(b) $\lim_{r \to \infty} e^{r\lambda_k} \exp r\, \nu_{n(k)} = 0$, for $k = 1, \ldots, m$.

and observe, that (a) implies (b) and thus $\lim_{r \to \infty} \pi(r) = 0$.

Now $\exp r\, \nu_{n(k)} = \sum_{p=o}^{n(k)} \frac{r^p}{p!} \nu_{n(k)}{}^p$. If e_i, $\hat{e}_i$ are dual bases of the k-th summand of V (respectively, of its dual), then

$$\langle \pi(r)\, e_i, e_j \rangle = e^{r\lambda_k} \sum_{p=o}^{n(k)} \delta_{i+p,j} = e^{r\lambda_k} \frac{r^{j-i}}{(j-i)!}$$

for $i \leq j$ and $= 0$ otherwise. Thus $R(S)$ is generated by the constants and the functions

$$F_{m,\lambda}(r) = \frac{r^m}{m!} e^{\lambda r}, \text{ Re } \lambda < 0, \ m = 0,1,2,\ldots$$

Thus R(S) is isomorphic to the algebra of all functions $r \longrightarrow c_1 r^{m_1} e^{\lambda_1 r} + \ldots + c_p r^{m_p} e^{\lambda_p r}$ with $c_g \in \mathbb{C}$, $\lambda_g \in \mathbb{C}$ and Re $\lambda_g < 0$.

Let us describe the characteristics of these examples coherently:

12.18. a) If N* denotes the one point compactification of the infinite cyclic semigroup $\{0,1,\ldots\}$ then all indecomposable identity preserving representations of N* are of the form

$$n \longmapsto (\lambda\, \epsilon_d + \nu_d)^n ,$$
$$\infty \longmapsto 0$$

where $\lambda \in \mathbb{C}$, $|\lambda| < 1$, ϵ_d is the identity of the d-dimensional representation space V and ν_d is a nilpotent endomorphism of V with $\nu_d e_p = e_{p+1}$ for a suitable basis $e_1,\ldots,e_d$, where $e_{d+1} = e_{d+2} = \ldots = 0$.

The representation algebra R(N*) is generated as an algebra by the constants and the functions

$$n \longmapsto \delta_{mn} \text{ (Kronecker delta) for } m = 0,1,2,\ldots$$

$$n \longmapsto \begin{cases} \binom{n}{m} \lambda^{n-m} & \text{for } m \leqq n \\ 0 & \text{otherwise.} \end{cases}$$

The semigroup $(N^*)' = \text{Spec } R(N^*)$ equals N^*.

(b) If H^* denotes the one point compactification of the additive half line $H = \{r \in R : r \leq 0\}$, then all indecomposable identity preserving representations of H^* are of the form

$$h \longmapsto \exp h\,(\lambda\, \epsilon_d + \nu_d),$$

where $\lambda \in \mathbb{C}$, Re $\lambda < 1$, and where ϵ_d, ν_d are as in (a).

The representation algebra $R(H^*)$ consists of all functions

$$h \longmapsto c_1\, h^{m_1}\, e^{\lambda_1 h} + \ldots + c_p\, h^{m_p}\, e^{\lambda_p h}$$

with c_g, $\lambda_g \in \mathbb{C}$, Re $\lambda_g < 0$, $m_g = 0, 1, 2, \ldots$.

The semigroup $(H^*)' = \text{Spec } R(H'^*)$ equals $R^* = R \cup \{+\infty\} = -H \cup H^*$.

Example (b) above shows a case where $S \neq S'$. We are working now towards conditions which will ensure that $S = S'$.

12.19. LEMMA. Let A be an involutive commutative algebra with identity and $S = \text{Spec } A$. Let $a \longmapsto \hat{a} : A \longrightarrow \tilde{C}(S)$ be the Gelfand transform of A into the involutive algebra of all continuous functions on S. Let $C(S) \subseteq \tilde{C}(S)$ be the subalgebra of all bounded functions. If $\hat{A} \subseteq C(S)$, then S is compact and $(\hat{A})^- = C(S) = \tilde{C}(S)$.

Proof. Let βS be the Stone-Čech compactification of S. Since the restriction morphism $C(\beta S) \longrightarrow C(S)$ is an isomorphism, we may consider all $\hat{a}$ as elements of $C(\beta S)$. If $s \in \beta S \setminus S$, then $a \longmapsto \hat{a}(s)$ is a morphism $A \longrightarrow \mathbb{C}$ of involutive algebras. Hence H is an element of $S = \text{Spec } A$ with $s(a) = \hat{a}(s)$. But we assumed $s \notin S$. This contradiction shows $S = \beta S$. Hence S is compact. Since $\hat{A}$ separates points, is involutive, and contains the constants, it is dense in $C(S)$ by the theorem of Stone and Weierstrass.

12.20. <u>LEMMA</u>. If, under the hypotheses of 12.19, there is a morphism φ of involutive algebras of A into a C*-algebra B, then there is a morphism of C*-algebras $\psi: C(S) \longrightarrow B$ such that

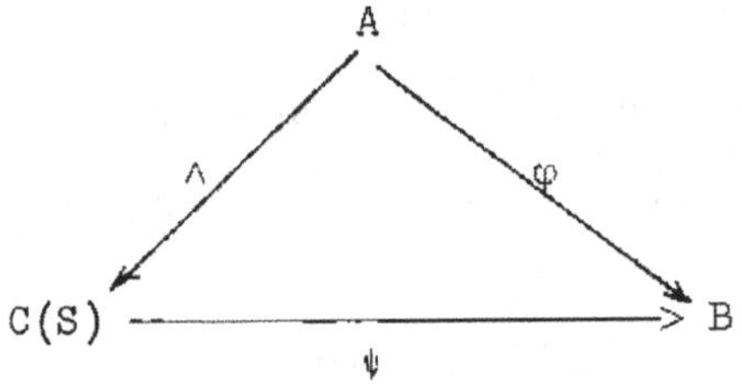

commutes.

Proof. The closure of $\varphi(A)$ in B is a commutative C*-algebra. We may therefore assume that B is commutative and $\varphi(A)$ is dense. Now $\text{Spec } \varphi: \text{Spec } B \longrightarrow S = \text{Spec } A$ is a continuous function, which is in fact injective since $\varphi(A)$ is dense in B. Let $\psi' = C(\text{Spec } \varphi): C(S) \longrightarrow C(\text{Spec } B)$ and let $i: C(\text{Spec } B) \longrightarrow B$ be the natural isomorphism. Then $\psi = i\psi'$ is the desired morphism.

12.21. PROPOSITION. *Let* $c: A \longrightarrow A \otimes^* A$ *be a commutative* C^*-*bigebra with identity. Then the following conditions are equivalent.*

(a) *The Gelfand transforms of all elements of* A' *are bounded.*

(b) Spec A' *is compact, and* W(Spec A) (12.13) *is epimorphically embedded in* Spec A'.

Proof. By 12.13 we may assume that c is saturated, i.e. that A' is dense in A and that Spec A may be identified with a subspace of Spec A'. If (a) is satisfied, then S' = Spec A' is compact by 12.19.

(b) ⟹ (a) is trivial.

12.22. LEMMA. Let S' be a Peter Weyl semigroup and G a compact subgroup of S' containing the identity, such that G is epimorphically embedded in any semigroup T with $G \subseteq T \subseteq S'$ relative to the category of Peter-Weyl semigroups. Then G is a maximal compact subgroup of H(1), the group of units of S', and G is its own normalizer in H(1).

Proof. As Poguntke has shown recently [14], epimorphisms in the category of compact groups are surjective. Thus, in view of 12.18, if K is a compact group with $G \subseteq K \subseteq S'$ then G = K. If N is the normalizer of G in H(1), then the quotient morphism $N \longrightarrow N/G$ and the constant morphism $N \longrightarrow N/G$ agree on G. Since G is epimorphically embedded in B (as N, being linear, is a Peter-Weyl group), G = N follows.

12.23. LEMMA. Under the hypotheses of 12.22 if $G \subseteq K \subsetneqq S'$ and K is a compact semigroup, then G = K.

Proof. In K the identity has a neighborhood which does not contain any idempotents different from 1. There are two cases: (a) G is open in K, (b) G is not open in K. In case (a) $K\backslash G$ is empty or is an open closed ideal. The constant representation $\rho: K \longrightarrow \text{End } V$ with $\rho(k) = 1$ and the representation $\pi: K \longrightarrow \text{End } V$ defined by $\pi(g) = 1$ for $g \in G$ and $\pi(K) = 0$ for $k \in K\backslash G$, if $G \neq K$, agree on G. This contradiction shows K = G in this case. In case (b) there is a one parameter semigroup $f: H \longrightarrow K$ with $f(0) = 1$, $f(H) \cap G = \{1\}$ such that $f(H)^*$ is in the centralizer of G ([11], p. 95). Then $G \longrightarrow G\, f(H)^*$ is an epic of Peter-Weyl semigroups. Now $G\, f(H)^*$ is a linear cylindrical semigroup ([11], p. 83 ff) and thus there is surjective homomorphism $\varphi: G\, f(H)^* \longrightarrow [0, 1]$ mapping G onto 1. (Note that all $f(r)$, $r \in H$ are units in End V, i.e. are invertible.) Let α be a non-trivial automorphism of [0, 1] (e.g. $\alpha(r) = r^2$). Then φ, $x\varphi$ are different morphisms of Peter-Weyl semigroups which agree on G. This is a contradiction to the epimorphy $G \longrightarrow G\, f(H)^*$. Thus case (b) is impossible, and the lemma is proved.

NOTE: By 12.19, S' cannot have any compact ideals; in particular S' has no zero.

12.24. COROLLARY. If G is a compact group and the Gelfand transforms of all $f \in R(G)$ are bounded, then $G = \text{Spec } R(G)$.

Proof. By 12.21, Spec $R(G)$ is compact and $G = W(G)$ is epimorphically embedded in Spec $R(G)$. Then 12.23 proves the remainder.

12.25. LEMMA. Let G be a compact group. Then the Gelfand transform of any $f \in R(G)$ is bounded.

Proof. The finite dimensional complex vector space $\langle f \rangle$ is a Hilbert space relation to the inner product $(g|h) = \int g\bar{h}\, d\mu$ with normalized Haar measure μ on G. Let $e_1,\ldots,e_n \in \langle f \rangle$ be an orthonormal basis. For $g \in \langle f \rangle$ we then have $g = \Sigma(g|e_i)e_i$. If $s \in G$ and $(s \cdot f)(x) = f(xs)$, then we have $f(xs) = \Sigma\, F_i(s)e_i(x)$ with $F_i(s) = (s \cdot f|e_i)$. Note $F_i \in R(S)$, and $f(s) = \Sigma\, F_i(s)e_i(1)$. Now $(f|f) = (s \cdot f|s \cdot f) = \Sigma\, F_i(s)\overline{F}_i(s)$. Let $m \in$ Spec $R(G)$ be a multiplicative hermitean functional. Then $(f|f) = m(\Sigma\, F_i\overline{F}_i) = \Sigma\, m(F_i)m(\overline{F}_i)$ and $|m(f)|^2 = m(f\bar{f}) = m(\Sigma\, F_i\overline{F}_k e_i(1)\, \overline{e_k(1)}) = \Sigma\, m(F_i)\, \overline{m(F_k)} e_i(1)\, \overline{e_k(1)}$. Let T be the hermitean operator on $\mathbb{C}^n$ with coefficient matrix $(e_i(1)\, \overline{e_k(1)})_{i,k=1,\ldots,n}$; let $M \in \mathbb{C}^m$ be the vector with components $m(F_i)$, and denote the standard inner product of $\mathbb{C}^n$ by $[\ |\ ]$. Then $[M|M] = (f|f)$ and $[M|TM] = |m(f)|^2$, whence $|m(f)| \leqq \|T\|^{1/2}\, (f|f)^{1/2}$.

Since T depends only on f, the function $m \longmapsto m(f)$ is bounded.

12.26. PROPOSITION. Let G be a compact group. Then Spec $R(G) = G$, i.e. the point evaluations on $R(G)$ are exactly the characters.

Proof. 12.24, 12.25.

REMARK. A more direct proof of 12.26 is given in [8]; but we use some purely structure theoretical lemmas which are of independent interest (12.22, 12.23).

Recall that 12.26 is in sharp contrast with the situation in semigroups, even in simple cases (see 12.18b).

As we observed in Sections 10 and 11, primitive elements play an important role in the theory of cogebras. The following discussion rounds this topic off as far as the general theory is concerned.

We first define the concept of a reduced semigroup algebra. Recall that a semigroup is *involutive* if there is an involutive selfmap $s \longrightarrow s^*$ with $(st)^* = t^*s^*$. Every abelian semigroup is involutive relative to the trivial involution $s^* = s$. Every group is involutive relative to the involution $s \longrightarrow s^{-1}$. Note that in Clifford semigroups this involution in each subgroup defines a global involution.

12.29. DEFINITION. Let S be a semigroup with zero $\overline{0}$ and $\mathbb{C}[S]$ the semigroup algebra of S over $\mathbb{C}$. *The reduced semigroup algebra* $\mathbb{C}_{red}[S]$ is $\mathbb{C}[S]/\mathbb{C}\cdot\overline{0}$, the algebra obtained by identifying $\overline{0}$ with the zero of the algebra. Clearly $\mathbb{C}_{red}[S]$ may be considered to be the vector space spanned by $S\backslash\{\overline{0}\}$ with the obvious multiplication. If S is involutive then both $\mathbb{C}[S]$ and $\mathbb{C}_{red}[S]$ are involutive relative to the involution defined by $(\Sigma\, a_s \cdot s)^* = \Sigma\, \overline{a_s} \cdot s^* = \Sigma\, \overline{a_{s^*}} \cdot s$.

We briefly recall that for a C*-algebra A the spectrum Spec A is the space of all (unitary equivalence classes of) irreducible representations. We define Spec_a A to mean the subspace of all one dimensional nontrivial representations. We may identify Spec_a A with the set of all nonzero C*-algebra morphisms $f:A \longrightarrow \mathbb{C}$, i.e. all characters. We have $\mathrm{Spec}_a A = \mathrm{Spec}\, A$ iff A is abelian. For an arbitrary involutive algebra A the space Spec_a A will be the set of all non-zero morphisms $A \longrightarrow \mathbb{C}$ of involutive algebras with the topology of point-wise convergence.

Most applications of the following theorem will be concerned with the commutative case, in which we may read Spec in place of Spec_a. We use the notation of 10.17.

12.30. THEOREM. *Let* $c:A \longrightarrow A \otimes^* A$ *be a* C*-*cogebra with identity* A_p' *the linear span of* P(A), *the involutive semigroup of primitive elements.*

(i) $c(A_p') \subseteq A_p' \otimes A_p'$, *i.e.* $A_p' \subseteq A'$ *and* A_p' *is an* Inv-cogebra,

(ii) A_p' *is the reduced involutive semigroup algebra* $\mathbb{C}_{red}[P(A)]$ *of* P(A), *and its comultiplication is the natural comultiplication of semigroup algebras.*

The function $\kappa':\mathrm{Spec}\, A_p' \longrightarrow \mathrm{Hom}(P(A),\mathbb{C}^\times)$ *with* $\kappa'(f) = f|P(A)$ *into the semigroup of involution, identity and zero preserving (not necessarily continuous) semigroup morphisms into the multiplicative semigroup of complex numbers is an isomorphism of topological semigroups if the range is given the topology of pointwise convergence.*

(iii) *The closure* A_p *of* A_p' *in* A *is a* C^*-*subalgebra of* A. *The function* $\kappa\colon \operatorname{Spec} A_p \longrightarrow \operatorname{Hom}(P(A),D)$ *with* $\kappa(f) = f|P(A)$ *into the semigroup of involution, identity and zero preserving (not necessarily continuous) non-zero semigroup morphisms into the complex unit disc* D *is an embedding of topological semigroups if the range is given the topology of pointwise convergence on the range. The diagram*

$$
\begin{array}{ccccccc}
\operatorname{Spec} A & \twoheadrightarrow & \operatorname{Spec} \overline{A'} & \twoheadrightarrow & \operatorname{Spec} A_p & \xrightarrow{\subset} & \operatorname{Hom}(P(A),D) \\
 & & \downarrow \cap & & \downarrow \cap & & \downarrow \cap \\
 & & \operatorname{Spec} A' & \longrightarrow & \operatorname{Spec} A_p' & \xrightarrow[\cong]{} & \operatorname{Hom}(P(A),\mathbb{C}^{\times})
\end{array}
$$

commutes. The top line consists of compact semigroups, all maps marked $\twoheadrightarrow$ *are quotient morphisms, all maps marked* $\subset$ *are embeddings.*

REMARK. If the Gelfand transforms of all elements of $P(A) \subseteq A_p'$ are bounded, then the Gelfand transforms of all elements of A_p' are bounded. Hence, by 12.19, $\operatorname{Spec} A_p'$ is compact. Then $\operatorname{Hom}(P(A),\mathbb{C}^{\times})$ is compact, which implies that $\overline{\varphi(P(A))}$ is compact for each $\varphi \in \operatorname{Hom}(P(A),\mathbb{C}^{\times})$. Since D is the largest compact subsemigroup of $\mathbb{C}^{\times}$, we have $\varphi \in \operatorname{Hom}(P(A),D)$. Thus $\operatorname{Hom}(P(A),D) = \operatorname{Hom}(P(A),\mathbb{C}^{\times})$ in this case. It is noteworthy that the zero morphism $P(A) \longrightarrow \mathbb{C}^{\times}$ in this case must be isolated in the set of *all* involutive morphisms $P(A) \longrightarrow D$ relative to the pointwise topology.

Proof. (i) follows from $c(a) = a \otimes a \in A_p' \otimes A_p'$ for all $a \in P(A)$.

(ii) That A_p' is the semigroup algebra with the asserted comultiplication is a consequence of 10.18 (ii) and the fact that $c(a) = a \otimes a$ for $a \in P(A)$.

Clearly $\kappa'(f)$ is an involution and zero preserving non-zero morphism of semigroups, and $\kappa'(f) = \kappa'(g)$ implies $f = g$ since $P(A)$ spans A_p'. If $\varphi: P(A) \longrightarrow \mathbb{C}^X$ is an involution, identity and zero preserving nonvanishing morphism, then the linear extension $f: A_p' \longrightarrow \mathbb{C}$ is a nonzero morphism of involutive $\mathbb{C}$-algebras such that $\kappa'(f) = \varphi$. Thus κ' is bijective. If $f, g \in \operatorname{Spec} A_p'$ then the product fg in the semigroup $\operatorname{Spec} A_p'$ is defined by $(fg)(a) = (f \otimes g)\, c(a)$. If $a \in P(A)$, then $\kappa'(fg)(a) = (f \otimes g)(a \otimes a) = f(a)\, g(a) = (\kappa'(f)\, \kappa'(g))(a)$. Thus κ' is a morphism of semigroups. Let f_i be a net in $\operatorname{Spec} A_p'$. This net converges to $f \in \operatorname{Spec} A_p'$ iff $f(a) = \lim f_i(a)$ for all $a \in A_p'$. Since a is a finite linear combination $\sum_{k=1}^{n} c_k p_k$ of primitive elements $p_k \in P(A)$ this is the case iff $f(p) = \lim f_i(p)$ for all $p \in P(A)$. But this in turn is equivalent to $\kappa'(f) = \lim \kappa'(f_i)$.

(iii) That A_p is a C*-cogebra in A is straightforward. If $f \in \operatorname{Spec} A_p$, then $|f(a)| \leqq ||a||$, whence $f(P(A)) \subseteqq D$. Thus κ is well defined. Since f induces an element $f|A_p' \in \operatorname{Spec} A_p'$, and $\kappa(f) = \kappa'(f|A_p')$, it follows that is a continuous injective semigroup morphism. Because of the compactness of the algebra spectrum $\operatorname{Spec} A_p$ we know that is a homeomorphism onto its image, i.e. an embedding.

The commutativity of the diagram is clear.

13. Dualities for compact groups

In this section we propose to show how the classical duality theorem for the category of compact groups and the category of compact abelian groups fits into our general duality theory. For the sake of completeness let us recall 11.6 and 11.7 and amplify slightly as follows:

13.1. PROPOSITION. *Let* $c : A \longrightarrow A \otimes^* A$ *be a proper commutative* C**-bigebra. Then the following are equivalent:*

(a) Spec A *is a compact group.*

(b) Spec A *is a group.*

(c) A *is right and left cosimple and has an identity.*

(d) A *is right cosimple and has an identity and a left coidentity.*

(e) A *is cosimple and has an identity and a coidentity.*

(f) A *has an identity* $1 : A \longrightarrow \mathbb{C}$, *a coidentity* $\epsilon : A \longrightarrow \mathbb{C}$, *and the* C**-algebra* A *has an automorphism* σ *of order* 2 *such that the diagram*

$$\begin{array}{ccccc} A & \xrightarrow{\ \epsilon\ } & \mathbb{C} & \xrightarrow{\ 1\ } & A \\ {\scriptstyle c}\downarrow & & & & \downarrow{\scriptstyle m} \\ A \otimes^* A & & \xrightarrow[\sigma \otimes^* A]{} & & A \otimes^* A \end{array}$$

commutes.

Moreover, if these conditions are satisfied, then the diagram in (f) *also commutes when* $\sigma \otimes^* A$ *is replaced by*

$A \otimes^* \sigma$, and the diagram

$$\begin{array}{ccc} A & \xrightarrow{\ \ \sigma\ \ } & A \\ {\scriptstyle c}\downarrow & & \downarrow{\scriptstyle c} \\ A \otimes^* A & \xrightarrow[\alpha(\sigma \otimes^* \sigma)]{} & A \otimes^* A \end{array}$$

commutes where α is the automorphism characterized by $\alpha(a \otimes b) = b \otimes a$. The full category of C-bigebras satisfying these conditions is dual to the category of compact groups.*

Proof. After the results of Section 11.6, all we have to do now is to investigate condition (f). Let $S = \text{Spec } A$. The dual of the morphism σ is an involution of the space S. The dual of the diagram in (f) is

$$\text{(S)} \qquad \begin{array}{ccccc} S \times S & & \xrightarrow{\ \ \hat{\sigma} \times S\ \ } & & S \times S \\ {\scriptstyle \Delta}\uparrow & & & & \downarrow{\scriptstyle m} \\ S & \longrightarrow & \{1\} & \xrightarrow{\ \subset\ } & S \end{array}$$

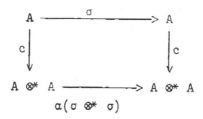

which means that $\hat{\sigma}(s)s = 1$ for all $s \in S$. Thus each element of the compact semigroups S has a left inverse, which implies that S is a group ([11], p. 18). Conversely, if S is a group, then inversion defines a space involution making the diagram (S) above commutative, when then results in the commutativity of the dual diagram in (f). Since in a group any left inverse is a right inverse, diagram (S) commutes with $S \times \hat{\sigma}$ replacing $\hat{\sigma} \times S$. Moreover, inversion in a group is an antiautomorphism. Dualizing these observations yield the last remark.

There is yet another characterization which is the dual of the following well known fact:

13.2. Let S be a compact semigroup. Let μ be a positive right subinvariant measure on S (i.e. $\mu\{s \in S : st \in X\} \leq \mu(X)$ for all Borel sets and all $t \in S$). Then the following are equivalent:

(a) S has a left identity and the support of μ is S.

(b) μ is also left subinvariant and the support of μ is S.

(c) S is a group and μ is a Haar measure.

([1], p. 88 ff.)

13.3. Let $c: A \longrightarrow A \otimes^* A$ be a C^*-cogebra, and $f: A \longrightarrow \mathbb{C}$ any C^*-morphism (character). Then there is a unique C^*-endomorphism $F_f: A \longrightarrow A$ such that the diagram

$$\begin{array}{ccc} A & \xrightarrow{\quad c \quad} & A \otimes^* A \\ {\scriptstyle F_f}\Big\downarrow & & \Big\downarrow{\scriptstyle A \otimes^* f} \\ A & \xrightarrow[\cong]{\qquad} & A \otimes^* \mathbb{C} \end{array}$$

commutes with the standard isomorphism at the bottom. For two characters $f, g: A \longrightarrow \mathbb{C}$, as we recall, we obtain a new character $fg: A \longrightarrow \mathbb{C}$ as the composition of

$$A \xrightarrow{c} A \otimes^* A \xrightarrow{f \otimes^* g} \mathbb{C} \otimes^* \mathbb{C} \xrightarrow{\cong} \mathbb{C}.$$

The identity $F_g F_f = F_{fg}$ then follows from the commutative diagram

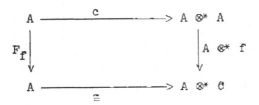

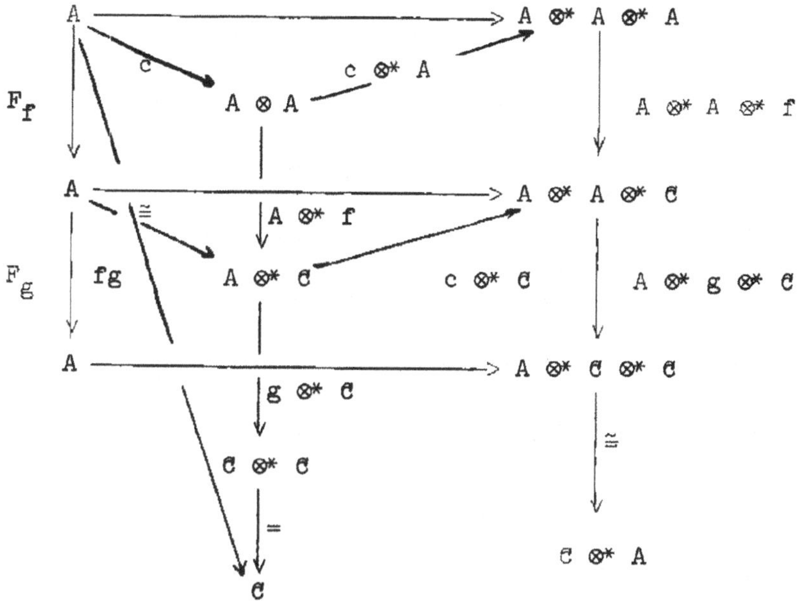

We write $f \cdot a$ for $F_f(a)$ and thus have $g \cdot (f \cdot a) = (gf) \cdot a$. The semigroup $C^*(A, \mathbb{C})$ (see 10.16) thus operates on A as a semigroup of C*-algebra endomorphisms. In other words $f \longmapsto F_f : C^*(A, \mathbb{C}) \longrightarrow \text{End } A$ is a morphism. Let now p be a positive functional on the C*-algebra A. We say that p is <u>right subinvariant</u>, if $p(f \cdot a) \leq p(a)$ for all positive $a \in A$, and all characters f of A. We call p <u>right-invariant</u> if $p(f \cdot a) = p(a)$ for all $a \in A$, $f \in \text{Spec } A$.

The concept of <u>left-subinvariance</u> and <u>left-invariance</u> is defined dually, and p is <u>subinvariant</u>, if b is both left and right subinvariant.

The following is an equivalent formulation of the equivariance of a functional p:

(a) Let $c: A \longrightarrow A \otimes^* A$ be a C*-cogebra. A positive functional is right invariant iff the following diagram commutes

$$\begin{array}{ccc} A & \xrightarrow{\quad c \quad} & A \otimes^* A \\ {\scriptstyle p}\downarrow & & \downarrow{\scriptstyle p \otimes^* s} \\ \mathbb{C} & \xrightarrow[\cong]{\qquad} & \mathbb{C} \otimes \mathbb{C} \end{array}$$

for each $s \in C^*(A, \mathbb{C})$, $s \neq 0$.

(b) Suppose that A has an identity and is commutative and semisimple, then p is right invariant iff

$$\begin{array}{ccc} A & \xrightarrow{\quad c \quad} & A \otimes^* A \\ {\scriptstyle p}\downarrow & & \downarrow{\scriptstyle p \otimes^* A} \\ \mathbb{C} & \xrightarrow{\qquad} & \mathbb{C} \otimes^* A \end{array}$$

commutes.

Proof. Statement (a) is a reformulation of the definition.

If A is commutative and semisimple, then the characters separate points. The assertion (b) then follows from (a).

The significance of this concept is best observed in the case of commutative C*-bigebras with identity.

13.4. Let $c: A \longrightarrow A \otimes^* A$ be a commutative C*-bigebra with identity. Let S be the dual semigroup. Then any positive linear functional of A corresponds to a positive Borel measure on S via $p(f) = \int f d\mu$ (if A and C(S) are identified). The functional p is right subinvariant if and only if the measure μ is right subinvariant.

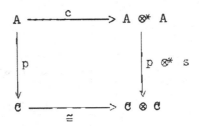

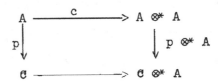

Proof. We may identify A with C(S). By the Riesz representation theorem a positive linear functional on A is given by a positive Borel measure μ and $p(f) = \int f d\mu$ for $f \in C(S)$. The comultiplication c under our identification may be considered as the morphism $c:C(S) \longrightarrow C(S \times S)$ defined by $c(f)(s, t) = f(st)$. We may also equalize $A \otimes^* \mathbb{C}$ with C(S). For $s \in S$, the morphism $A \otimes^* s:A \otimes^* A \longrightarrow A \otimes^* \mathbb{C}$ is then identified with the function mapping a function $f:S \times S \longrightarrow \mathbb{C}$ into the function $t \longrightarrow f(ts)$; because for $g, h \in C(S)$ we have $(A \otimes^* s)(g \otimes^* h) = g \otimes h(s)$ so that if we interpret $g \otimes h$ as the function $(x, y) \longrightarrow g(x)\, h(y)$, we have $[A \otimes^* s)(g \otimes^* h)](t) = g(t)h(s)$; since $A \otimes A$ is dense in $A \otimes^* A$ our assertion follows.

With the present identifications, for $f \in A$ and $s \in S$ we have $(s \cdot f)(t) = f(st)$. Thus $p(s \cdot f) \leq p(f)$ is equivalent to $\int (s \cdot f)\, d\mu \leq \int f\, d\mu$. This means that the measure μ is right subinvariant.

13.5. <u>COROLLARY</u>. On a commutative C*-bigebra with identity there exist right subinvariant positive functionals if and only if the unique maximal bideal is a maximal left bideal.

Proof. This is the dual statement to the following fact known in the theory of compact semigroups: A compact semigroup has a right subinvariant measure if and only if its minimal ideal is a minimal left ideal. (See e.g. [1], p. 97.)

This corollary indicates the simplest example of a C*-bigebra without right invariant positive functional: Let S be any compact Space, $A = C(S)$ and define $c: A \longrightarrow A \otimes^* A = C(S \times S)$ via $c(f)(x, y) = f(y)$. Then $s \cdot f$ is the constant function $t \longrightarrow f(s)$, whence $\int (s \cdot f) d\mu = f(s)\mu(S)$ for each Borel measure μ and each $f \in C(S)$. It is then readily seen directly, that no right subinvariant measure exists. Of course this follows from the general theorem mentioned in the proof of 13.5, since we made S into a right zero semigroup, which is simple, but not left simple, since all singleton sets are left ideals.

The semigroup result quoted previously has in fact the following corollary:

13.6. COROLLARY. Let S be a compact semigroup. The following statements are equivalent:

(i) S is a group.

(ii) S has a positive subinvariant measure with full support (i.e. support μ = S).

(iii) S has a positive right subinvariant measure with full support and possesses a left identity.

(See [1], p. 100.)

The dual of this assertion then immediately gives us the following complement to Proposition 13.1:

13.7. COMPLEMENT. Under the hypotheses of 13.1, the following conditions are equivalent to any of the other conditions in 13.1:

(g) A has a positive subinvariant functional p with

the property that $p(a^*a) = 0$ *implies* $a = 0$.

(h) A *has a positive right invariant functional with the property that* $p(a^*a) = 0$ *implies* $a = 0$, *and* A *has a right coidentity*.

After we have characterized the duals of groups satisfactorily in the category $bi[C^*_a]$, the question is to what extent the duals of groups may be characterized in the category bi[Inv].

By Proposition 12.26 we know that for any compact group S and its dual C*-cogebra $C:A \longrightarrow A \otimes^* A$ the algebras $A = C(S)$ and $A' = R(S) \subseteqq C(S)$ have the same spectrum. Let p' be the restriction of p to A'. Then p' has the following properties:

(1) p' is a strictly positive functional (i.e. $p'(a^*a) \geqq 0$ and equality holds iff $a = 0$).

(2) p' is right invariant (i.e. $p'(s \cdot a) = p(a)$ for $a \in A'$, $s \in S$) and left invariant.

<u>REMARK</u>. The discussion 13.3 holds in Inv just as well as in C*. Thus (2) is perfectly well defined. Conversely, suppose that $c':A' \longrightarrow A' \otimes A'$ is an Inv-cogebra with a left coidentity ϵ and that p' is a strictly positive invariant functional. Then Spec A' is compact, as follows through a proof modelled after and generalizing the proof of 12.25: Let $a' \in A'$, then $c'(a') = \sum_{i=1}^{n} e_i \otimes a_i$; the vector subspace of A' spanned by $e_1, \ldots, e_n$ is a Hilbert space relative to the inner product $(a|b) = p(b^*a)$; we may assume that the e_i are an orthonormal basis. Also

$c'(a'^*a') = \Sigma\, e_i^*\, e_k \otimes a_i^*\, a_k$ whence $(p \otimes s)\, c'(a'^*a) =$ $\Sigma\, p(e_i^*\, e_k) \otimes \overline{s(a_i)}\, s(a_k) = 1 \otimes \Sigma\, \overline{s(a_i)}\, s(a_i)$ for all $s \in \text{Spec } A'$. By right invariance, i.e. the commutativity of the diagram

$$\begin{array}{ccc} A' & \xrightarrow{\;c'\;} & A' \otimes A' \\ {\scriptstyle p}\downarrow & & \downarrow{\scriptstyle p \otimes s} \\ \mathbb{C} & \xrightarrow{\;\cong\;} & \mathbb{C} \otimes \mathbb{C} \end{array}$$

we have then $p(a'^*a') = \Sigma\, s(a_i)\, \overline{s(a_i)}$. Since ϵ is a left coidentity, we also get $a'^*a' = \Sigma\, \epsilon(e_i^*\, e_k)\, a_i^* a_k =$ $\Sigma\, g_{ik}\, a_i^*\, a_k$ with $g_{ik} = \overline{\epsilon(e_i)}\, \epsilon(e_k)$ as the coefficients of a hermitean operator T. Thus $|s(a)|^2 = s(a^*a) =$ $\Sigma\, g_{ik}\, \overline{s(a_i)}\, s(a_k) \leqq ||T||^2(\Sigma\, s(a_i)\, \overline{s(a_i)})^{1/2} =$ $||T||\, p(a'^*a')$ and $||T||$ does not depend on s.

Thus we have the following theorem:

13.8. *THEOREM. Let* $c: A \longrightarrow A \otimes A$ *be a commutative* Inv-*cogebra and* S *the topological semigroup* Spec A. *Then the following statements are equivalent.*

(a) S *is a compact group.*

(b) c *has a left coidentity and* A *has a right invariant strictly positive functional.*

If these conditions are satisfied, and A *is in addition semisimple, then* c *is naturally isomorphic to the cogebra* $R(S) \longrightarrow R(S) \otimes R(S)$.

Proof. The last assertion must be proved. We may identify A with a dense subalgebra of C(S). By $C(S)' = R(S)$ we have

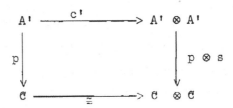

$A \subseteq R(S)$. As an S-module under $(s \cdot f)(x) = f(xs)$ the vector space $R(S)$ is semisimple in the sense that it is a direct sum of finite dimensional simple S-modules: For let $\{E_i : i \in I\}$ be a maximal family of simple finite dimensional orthogonal submodules (relative to the pre-Hilbert space structure induced by $L^2(G)$). Then $\Sigma E_i \subseteq R(S)$. Let $f \in R(S) \setminus \Sigma E_i$. We may assume that f is orthogonal to ΣE_i. Now $\langle f \rangle$ is a finite dimensional S-module, which is semisimple since S is a compact group; it is therefore a unique direct sum of simple submodules not all of which can be contained in ΣE_i, since $f \notin \Sigma E_i$. Let $F \not\subseteq \Sigma E_i$ be one of them. Since $f \perp \Sigma E_i$, we have $F \perp \Sigma E_i$. This contradicts the maximality of $\{E_i : i \in I\}$. So $\Sigma E_i = R(G)$. Now A is a submodule of $R(G)$. The same argument applies to A. We may therefore assume that $A = \Sigma \{E_i : i \in J\}$, $J \subseteq I$. Suppose $i_0 \in I \setminus J$, $f \in E_{i_0}$, $f \neq 0$. Since A is dense in $R(G)$, for each $\epsilon > 0$ there is a $g \in A$ with $\epsilon > (f - g | f - g) = (f|f) + (g|g)$ since $f \perp A$. With $\epsilon = \frac{1}{2}(f|f)$ we obtain a contradiction. Thus $A = R(G)$ as asserted.

The following is just a reinterpretation of the previous results; it is the so called TANNAKA DUALITY THEOREM.

13.9. THEOREM. *The full category of commutative semisimple Inv-cogebras with left coidentity and a right invariant positive functional is dual to the category of compact groups. The duality is given by the cofunctors*

Spec and R.

Later we will see how the Pontryagin duality fits into this picture.

14. Dualities for totally disconnected compact semigroups

In this section we will indicate how the category of compact totally disconnected semigroups which we briefly encountered in the context of Proposition 11.8, may be characterized further in terms of duality.

Let S be a compact totally disconnected semigroup with identity. It is then a projective limit of finite semigroups and is, consequently, a Peter-Weyl semigroup; indeed, any finite semigroup F with identity has a faithful linear representation, namely the one induced by left multiplication on the semigroup algebra $\mathbb{C}[F]$; the identity is essential here; note that some finite semigroups still give rise to a faithful left regular representation even if they do not have identities; this holds e.g. for all semilattices.

Any continuous representation π of a compact semigroup S with finite image is locally constant. On a totally disconnected compact semigroup with identity the continuous representations with finite image separate the points. Thus the algebra $R_{loc}(S)$ of locally constant functions on such an S is a point separating subalgebra of C(S). Moreover, as the dual of the semigroup multiplication, the comultiplication of C(S) induces a comultiplication on

$R_{loc}(S)$ making it an Inv-bigebra.

Conversely, let $f \in C(S)$ be locally constant. Then S is the disjoint union of the open closed sets U_i, $i = 1,\ldots,n$ on which f is constant. Thus $V = \cup\{U_i \times U_i : i = 1,\ldots,n\}$ is a neighborhood of the diagonal in $S \times S$. Since the set of congruences with open closed cosets has this diagonal as intersection, and since S is compact there is a congruence ρ on S with $\rho \subseteq V$. Thus any coset of ρ is contained in some U_j and f is constant on the cosets of S. It therefore defines a function $\bar{f} \in C(S/\rho) = \mathbb{C}^{S/\rho}$. Since the representations of S/ρ separate the points, we have $R(S/\rho) = C(S/\rho) = \mathbb{C}^{S/\rho}$. Hence $\bar{f} \in R(S/\rho)$ which means that f is the coefficient of a finite dimensional, continuous linear representation. Thus $R_{loc}(S) \subset R(S)$. We may have $R_{loc}(S) \neq R(S)$ (12.18.a).

Conversely, if S is a compact Peter-Weyl semigroup such that $R_{loc}(S)$ separates points, then S must be totally disconnected, since a locally constant function is constant on connected components.

We thus have

14.1. PROPOSITION. *Let S be a compact semigroup with identity. Then the following statements are equivalent:*

(a) *S is a Peter-Weyl semigroup and the bigebra $R_{loc}(S)$ of all locally constant functions (i.e. the set of all continuous functions $S \longrightarrow \mathbb{C}_d$ where $\mathbb{C}_d$ is the discrete field of complex numbers) separates points.*

(b) S _is totally disconnected_.

We recall the following definition.

14.2. _DEFINITION_. An algebra is called _biregular_ if every principal ideal is generated by a central idempotent.

It is well known that every primitive ideal in a biregular ring A is maximal modular, and that the structure space Prim A is totally disconnected locally compact Hausdorff, and in fact compact if A has an identity. In case that A is in addition commutative and an algebra over $\mathbb{C}$ such that any maximal ideal I is an algebra ideal and the quotient morphism $A \longrightarrow A/I$ is an isomorphism of $\mathbb{C}$-algebras, then A is isomorphic to the algebra of all continuous functions from Prim A into the _discrete_ field of complex numbers. The spaces Prim A (with its hull kernel topology) and Spec A with the weak star topology may be identified in this case. (For a detailed discussion of biregular rings see e.g. Dauns, J. and K. H. Hofmann: The representation of biregular rings by sheaves, Math. Zeitschr. _91_ (1966), 103-123.) Proposition 14.1 says that for a compact totally disconnected semigroup S with identity $R_{loc}(S)$ is biregular and that $S = \text{Spec } R_{loc}(S)$. In particular, this implies $S' = S$ with the notation of 12.14. This observation yields a very general proof of the fact that $N^{*\prime} = N^*$ which was established with considerable technical effort in 12.18 (although detailed additional information was amassed in the process).

14.3. <u>DEFINITION</u>. An (involutive) bigebra will be called <u>biregular</u>, if the underlying algebra is biregular. We say that it has <u>small stalks</u>, if for any maximal modular ideal I of A the map $c \longmapsto c.1 : \mathbb{C} \longrightarrow A/I$ is an isomorphism of $\mathbb{C}$-algebras.

It is well known that the category of biregular $\mathbb{C}$-algebras with small stalks and identity is dual to the category of compact totally disconnected spaces, under the duality (Spec, R_{loc}) (where Spec A is the space of maximal modular ideals of A or, equivalently, the space of algebra morphisms $A \longrightarrow \mathbb{C}$, and where $R_{loc}(X)$ is the algebra of locally constant functions $X \longrightarrow \mathbb{C}$. Since relative to the algebraic tensor product both cofunctors are readily seen to be multiplicative, we obtain the following theorem:

14.4. <u>THEOREM</u>. <u>The categories of compact totally disconnected representable semigroups and the category of commutative biregular complex bigebras with identity and with small stalks are dual under the duality</u> (R_{loc},Spec), <u>and</u> $R_{loc}(S) \subset R(S)$.

<u>REMARK</u>. Recall that any compact totally disconnected semigroup with identity is representable and that any compact totally disconnected semilattice is representable.

If S is a compact totally disconnected semigroup which is either a group or a semilattice, then any finite dimensional continuous representation has a finite image, since any totally disconnected Lie group is discrete and any compact totally disconnected subsemilattice End V for a finite dimensional vector space V is finite. For these semigroups we may therefore conclude $R_{loc}(S) = R(S)$ (as opposed to Example 12.18.a). We may therefore formulate the following duality theorem:

14.5. THEOREM. Under the duality (R,Spec) the category of compact totally disconnected groups [resp. semilattices] is dual to the category of commutative complex biregular bigebras with identity which are in addition left and right cosimple [resp. cocommutative and idempotent].

The theory outlined in this section depends very little on the choice of the ground field. A particularly suitable alternative choice would be the field GF(2) of two elements. There is thus a theory of Boolean bigebras in sight appearing as the dual category of the category of totally disconnected compact semigroups.

15. Discrete involutive semigroups and duality

This section will show how Pontryagin duality and the duality theory of compact totally disconnected and discrete semilattices (which is much less known) emerge from the general theory of bigebras.

15.1. Let S be an involutive semigroup (i.e. a semigroup with an involutory map $s \longrightarrow s^*$ satisfying $(st)^* = t^*s^*$; see remarks preceeding 12.29). Let $A = \mathbb{C}[S]$ be the involutive semigroup algebra. Then A is in fact a cocommutative Inv-cogebra relative to the natural comultiplication $c:A \longrightarrow A \otimes A$ defined by $c(s) = s \otimes s$ for $s \in S$. It has a coidentity $e:A \longrightarrow \mathbb{C}$ defined by $e(s) = 1$ for all $s \in S$. The involutive semigroup $P(A)$ of primitive elements is exactly $S \cup \{0\}$.

Proof. Most of these assertions are quite straightforward. That the augmentation $e:A \longrightarrow \mathbb{C}$ is a coidentity needs to be checked only for the generators $s \in S$: We indeed have $(e \otimes A)c(s) = (e \otimes A)(s \otimes s) = 1 \otimes s$; further $\mathbb{C}$ may be considered as the semigroup algebra of the singleton semigroup $\{1\}$ and e is just the algebra morphism induced functorially by the constant morphism $S \longrightarrow \{1\}$. If $a \in P(A)$ and $a \neq 0$, then $a = \Sigma\, a_s \cdot a$ for suitable elements $a_s \in \mathbb{C}$. Then we have $\Sigma\, a_s a_t \cdot s \otimes t = a \otimes a = c(a) = \Sigma\, a_s \cdot c(s) = \Sigma\, a_s \cdot s \otimes s$, whence

$$a_s a_t = \begin{cases} a_s & \text{if} \quad s = t \\ 0 & \text{if} \quad s \neq t \end{cases} .$$

As was shown in the proof of 10.18 (ii) this implies that all a_s are zero or that exactly one of them is 1 and the others zero.

15.2. Under the hypotheses of 15.1, Spec A is a commutative completely regular topological semigroup with identity e.

Proof. This follows from 12.6 and the fact that A is cocommutative and e is a coidentity.

15.3. Let Hom $(S,\mathbb{C}^{\times})$ denote the topological semigroup of identity and involution preserving semigroup morphisms into the multiplicative semigroup of complex numbers with the topology of pointwise convergence. Let κ :Spec A $\longrightarrow$ Hom$(S,\mathbb{C}^\times)$ be defined by $\kappa(f) = f|S$. Then κ is an isomorphism of topological semigroups.

The proof is the same as the one for 12.30(ii).

We denote Spec A with $\hat{S}_1$.

15.4. Let $D = \{z \in \mathbb{C}: |z| \leq 1\}$. Then Hom$(S,D)$ is (essentially) a compact subsemigroup of Hom$(S,\mathbb{C}^\times)$. Then $\hat{S} = \kappa^{-1}$ (Hom(S,D)) is a compact subsemigroup of $\hat{S}_1$ containing the identity e.

As usual we denote with $a \longmapsto \hat{a}:A \longrightarrow \tilde{C}(\hat{S}_1)$ the Gelfand transformation into the involutive algebra of <u>all</u> continuous functions on $\hat{S}_1$. Then $a \longrightarrow \hat{a}|\hat{S}:A \longrightarrow C(\hat{S})$ is a morphism of involutive algebras. Since $\hat{S}$ is a compact commutative semigroup with identity, $C(\hat{S})$ is a commutative C*-bigebra with identity and coidentity. The involutive cocommutative Inv-bigebra $R(\hat{S})$ with identity and coidentity is then well defined according to section 12.

Let $\hat{A}$ be the image of A in $C(\hat{S})$ under $\hat{a} \longrightarrow \hat{a}|S$. Now $\hat{A} \otimes \hat{A} \subseteq C(\hat{S}) \otimes C(\hat{S}) \subseteq C(\hat{S}) \otimes^* C(\hat{S}) \cong C(\hat{S} \times \hat{S})$; thus $\hat{a}_1 \otimes \hat{a}_2 \in \hat{A} \otimes \hat{A}$ may be identified with a continuous function $\hat{S} \times \hat{S} \longrightarrow \mathbb{C}$ so that $(\hat{a}_1 \otimes \hat{a}_2)(f_1 \otimes f_2) = \hat{a}_1(f_1)\hat{a}_2(f_2) = f_1(a_1)f_2(a_2)$ with $a_i \in A$, $f_i \in \hat{S} \subseteq$ Spec A, $i = 1,2$. If $s \in S \subseteq A$, then $c(a) = a \otimes a$; thus, if $\hat{c}:C(\hat{S}) \longrightarrow C(\hat{S}) \otimes^* C(\hat{S}) = C(\hat{S} \times \hat{S})$ is the comultiplication of $C(\hat{S})$, then for $a \in S$ we have $(\hat{\ } \otimes \hat{\ })c(a) = \hat{a} \otimes \hat{a}$, and $(\hat{a} \otimes \hat{a})(f_1 \otimes f_2) = f_1(a)f_2(a) = f_1f_2(a) = \hat{a}(f_1f_2)$; on the other hand $\hat{c}(\hat{a})(f_1 \otimes f_2) = \hat{a}(f_1 f_2)$. This shows that $\hat{A}$ is an involutive subbigebra of $C(\hat{S})$ and that $\hat{}$ is a morphism of bigebras. We can now conclude $\hat{A} \subseteq R(\hat{S}) \subseteq C(\hat{S})$.

If $a \in P(A)$, then $\hat{a} \in P(\hat{A})$ since $\hat{c}(\hat{a}) = \hat{} \otimes \hat{}\ c(a)$ and $c(a) = a \otimes a$. Since A is spanned by $P(A)$ and $S' = P(\hat{A}) \setminus \{0\} \subseteq \hat{A}$ is linearly independent, we have $P(A)^{\hat{}} = P(\hat{A})$ and S' is the image of S under $a \longmapsto \hat{a}$; further $\hat{A} = \mathbb{C}[S']$. Note that S' is commutative and consists of a point-separating family of semicharacters of $\hat{S}$ via $\hat{s}(f) = f(s)$ for $s \in S$, $f \in \hat{S} \subseteq \text{Spec } A$. Hence $\hat{A}$ is dense in $C(\hat{S})$, thus $(\hat{A})^- = R(S)^- = C(\hat{S})$ and $\hat{S}$ is a saturated compact Peter-Weyl semigroup. The inclusions $\hat{A} \longrightarrow R(\hat{S}) \longrightarrow C(\hat{S})$ define morphisms of topological semigroups $\hat{S} \longrightarrow \hat{S}' \longrightarrow \hat{S}_1$ with $\hat{S}' = \text{Spec } R(\hat{S})$ as in Section 12. Since $\hat{S}$ has a coidentity, $P(R(\hat{S})) \setminus \{0\} = P(C(\hat{S})) \setminus \{0\}$ is an involutive semigroup $\hat{\hat{S}}$; its span in $R(S)$ is the semigroup algebra $\mathbb{C}[\hat{\hat{S}}]$, so that we have the inclusions $\hat{A} \longrightarrow \mathbb{C}[\hat{\hat{S}}] \longrightarrow R(\hat{S}) \longrightarrow C(\hat{S})$. We then have morphisms $S \longrightarrow S' \longrightarrow \hat{\hat{S}}$ of semigroups and morphisms of topological semigroups $\hat{S} \longrightarrow S' \longrightarrow \text{Spec } \mathbb{C}[\hat{\hat{S}}] \longrightarrow \text{Spec } A \cong \text{Spec } \mathbb{C}[S]$. The morphism $S \longrightarrow S'$ is surjective and $S' \longrightarrow \hat{\hat{S}}$ is injective as is $\hat{S} \longrightarrow \hat{S}'$.

15.5. THEOREM. *Let* S *be a (discrete) involutive semigroup,* $A = \mathbb{C}[S]$ *its semigroup* Inv-*cogebra,* $\hat{S}_1 = \text{Spec } A$ *its spectrum (which is completely regular semigroup, and whose identity is the coidentity* e *of* A*). Let* $\text{Hom}(S, \mathbb{C}^{\times})$ *be the semigroup of all nonzero involution preserving* *semigroup morphisms with the topology of pointwise convergence. Let* $\kappa:\hat{S}_1 \longrightarrow \mathrm{Hom}\,(S,\mathbb{C}^\times)$ *be the function defined by* $\kappa(f) = f|S$.

(a) κ *is an isomorphism.*

Let $\hat{S} = \kappa^{-1}\,\mathrm{Hom}(S,D)$, $D = \{z\colon |z| \le 1\}$. *Then* $\hat{S}$ *is a compact Peter-Weyl semigroup with identity* e. *The Gelfand transformation defines a morphism of* Inv-*cogebras* $a \longmapsto \hat{a}|\hat{S}\colon A \longrightarrow R(\hat{S})$. *The image* $\hat{A}$ *of* A *in* $R(\hat{S})$ *is dense in the* C*-*bigebra* $C(\hat{S})$. *We have a diagram*

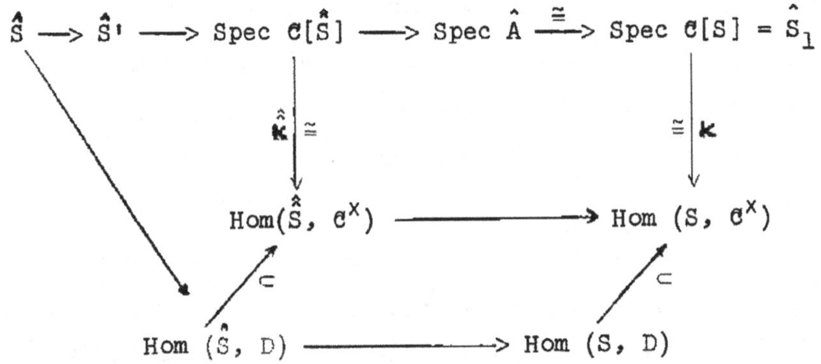

and the diagram commutes. In particular, $\hat{S} \cong \mathrm{Hom}\,(S,\,D)$ *is a retract of* $\mathrm{Hom}\,(\hat{\hat{S}},\,D)$.

Let us consider the simplest example (which complements our discussions around 12.18).

15.6. Let $S = N = \{0,1,2,\ldots\}$ the infinite cyclic semigroup of natural numbers. The semigroup algebra $A = \mathbb{C}[S]$ is the ring of polynomials $\mathbb{C}[X]$ in one variable over $\mathbb{C}$. The semigroup $S_1 = \mathrm{Spec}\,A$ of involutive multiplicative functionals is the semigroup $\mathbb{C}^\times$ whereby each $c \in \mathbb{C}$ defines the character $p(X) \longrightarrow p(c)$ for $p(X) \in \mathbb{C}[X]$. Obviously,

$\mathrm{Hom}(S,\mathbb{C}^\times)$ is also isomorphic to $\mathbb{C}^\times$ under the map which sends a semicharacter $\varphi:S \longrightarrow \mathbb{C}^\times$ onto $\varphi(1)$. Under this isomorphism, $\mathrm{Hom}(S,D)$ corresponds to D itself, so that we may identify $\hat{S}$ with D. The ring $\hat{A}$ is the ring of polynomial functions $\mathbb{C}[z]$ on the disc D, which is clearly dense in $C(\hat{S})$. The ring $R(\hat{S}) = R(D)$ contains all polynomials of $\mathbb{C}[z]$. But, as we will see now, $\mathbb{C}[z] \neq \mathbb{C}[\hat{\hat{S}}]$, hence in particular $\mathbb{C}[z] \neq R(D)$. The non-zero primitive elements of $R(D)$ are the semicharacters $D \longrightarrow D$ preserving the identity. Outside the constant semicharacter $z \longrightarrow 1$ they are of the form $z \longrightarrow |z|^\lambda z^n$, $n \in Z$, $\lambda \in \mathbb{C}$, $\mathrm{Re}\,\lambda > -n$. If one represents the non-zero $z \in D$ as $e^{-h+2\pi ir}$ with $h \in H = [0,\infty]$, $r + \mathbb{Z} \in \mathbb{R}/\mathbb{Z}$, then any such semicharacter is represented by

$$e^{-h+2\pi ir} \longmapsto e^{-ah+2\pi i(nr+bh)},\quad a,\ b \in R,\ a > 0,\ n \in Z$$
$$0 \longmapsto 0.$$

Thus $\hat{\hat{S}}$ is isomorphic to $(\mathbb{H}_o \times \mathbb{R} \times \mathbb{Z}) \cup \{(0,\ 0,\ 0)\} \subseteq R^3$ with $\mathbb{H}_o$ as the additive semigroup of positive reals. The involution on $\hat{\hat{S}}$ then corresponds to $(h,\ r,\ z) \mapsto (h,\ -r,-z)$, and the injection $S \longrightarrow \hat{\hat{S}}$ to $n \longmapsto (n,\ 0,\ n)$. Thus $\mathbb{C}[z] \neq \mathbb{C}[\hat{\hat{S}}]$. Then $\mathrm{Hom}(\hat{\hat{S}},\ \mathbb{C}^\times)$ is isomorphic to the semigroup of <u>all</u> involution preserving semigroup morphisms $\mathbb{H}_o \times \mathbb{R} \times \mathbb{Z} \longrightarrow \mathbb{C}^\times$ with the pointwise topology. The non-zero ones among them form a subsemigroup -- in fact a subgroup, namely $\mathrm{Hom}_{\mathrm{inv}}(\mathbb{H}_o \times \mathbb{R} \times \mathbb{Z},\ \mathbb{R} \times \mathbb{R}/\mathbb{Z}) \cong \mathrm{Hom}_{\mathrm{inv}}(\mathbb{R} \times \mathbb{R} \times \mathbb{Z},\ \mathbb{R} \times \mathbb{R}/\mathbb{Z})$ with $(r,\ t + Z) \longrightarrow (r,\ -t + \mathbb{Z})$ as involution on the range. This group is isomorphic to

$\mathrm{Hom}(\mathbb{R}, \mathbb{R}) \times \widehat{\mathbb{R}_d} \times \mathbb{R}/\mathbb{Z}$, where $\widehat{\mathbb{R}_d}$ is the character group of the discrete reals (the universal solenoidal group) and where $\mathrm{Hom}(\mathbb{R}, \mathbb{R})$ is isomorphic to the topological R-vector space $\mathbb{R}^X$ for some set X of continuum cardinality. The spectrum $\hat{S}'$ of $R(\hat{S})$ is the largest representable semigroup containing D such that all finite dimensional continuous representations of D extend uniquely to $\hat{S}'$. Clearly $\hat{S}'$ contains $\mathbb{C}^X$. The identity representation $D \longrightarrow D$ extends uniquely to a representation $S' \longrightarrow \mathbb{C}^X$, which is the identity when restricted to $\mathbb{C}^X$. Now the identity $S' \longrightarrow S'$ and the morphism $S' \longrightarrow \mathbb{C}^X \longrightarrow S'$ agree on D. Hence they agree by 12.16. Thus $S' = \mathbb{C}^X$. This is necessarily different from Spec $\mathbb{C}[\hat{\hat{S}}]$, hence $\mathbb{C}[\hat{\hat{S}}] \neq R(S)$. We thus have a sequence $\mathbb{C}[S] \subsetneq \mathbb{C}[\hat{\hat{S}}] \subsetneq R(\hat{S})$ of three different Inv-bigebras contained in the C*-bigebra $C(\hat{S})$, all of which are dense. [The inequality of $\mathbb{C}[\hat{\hat{S}}]$ and $R(\hat{S})$ also follows from the presence in $R(\hat{S})$ of the functions $z \longmapsto (\log|z|)^m z^\lambda$, $m \in N$, $\lambda \in \mathbb{C}$, $\mathrm{Re}\,\lambda > 0$, $0 \longmapsto 0$, which follows from 12.18. Thus we have the following table

S	$\hat{\hat{S}}$
$\mathbb{N}$	$(\mathbb{H}_0 \times \mathbb{R} \times \mathbb{Z}) \cup \{(0, 0, 0)\}$

$\hat{S}$	$\hat{S}'$	Spec $\mathbb{C}[\hat{\hat{S}}]$	$\hat{S}_1$
D	$\mathbb{C}^X$	$(\mathbb{R}^X \times \mathbb{R}_d \times \mathbb{R}/\mathbb{Z}) \cup \{0\}$	$\mathbb{C}^X$

$\mathbb{C}[S]$	$\mathbb{C}[\hat{\hat{S}}]$	$R[\hat{S}]$	$C(\hat{S})$
$\mathbb{C}[z]$	$\mathbb{C}[\|z\|^\lambda z^n]$, $n \in Z$, $\lambda \in \mathbb{C}$, $\mathrm{Re}\,\lambda > -n$	$R[D]$	$C(D)$

The example which we have just discussed illustrates the typical complications arising in the analysis of abelian semigroups even in very simple special cases. It is, therefore, noteworthy that groups and semilattices represent exceptions from this general rule.

Let us first discuss the case of groups. Let S be a (discrete) group with the natural involution $s \longrightarrow s^{-1}$, $\mathbb{C}[S]$ its semigroup Inv-cogebra. The spectrum $\hat{S}_1$ of $\mathbb{C}[S]$ is isomorphic to $\mathrm{Hom}(S, \mathbb{C}^{\times})$ the semigroup of involution and identity preserving semigroup morphisms. Since S is a group this semigroup is in fact the group $\mathrm{Hom}(S, \mathbb{C}^{\times}\backslash\{0\})$ relative to the pointwise topology whereby we observe that only involution preserving morphisms φ are admitted. This means $\varphi(s^{-1}) = \overline{\varphi(s)}$, hence $1 = \varphi(1) = \varphi(s)\varphi(s^{-1}) = |\varphi(s)|^2$. Thus, in fact we have $\hat{S}_1 \cong \mathrm{Hom}(S, \mathbb{R}/\mathbb{Z})$; this is a compact group, the ordinary character group, and in the notation of Theorem 15.5 we have $\hat{S} = \hat{S}_1$. Thus the spectrum of the commutative Inv-bigebra $\mathbb{C}[S]$ is a compact group. By 13.8 the latter is therefore isomorphic to the Inv-bigebra $R(\hat{S})$. In particular, the semigroup S of nonzero primitive elements of $\mathbb{C}[S]$ is isomorphic to the semigroup $\hat{\hat{S}}$ of nonzero primitive elements of $R(\hat{S})$ under the Gelfand representation. Thus $\mathbb{C}[S]$ is isomorphic to the linear span $C[\hat{\hat{S}}]$ in $R(\hat{S})$. Since $R(\hat{S})$ is the algebra of all coefficients of finite dimensional linear representations of the compact group $\hat{S}$ and all such are completely reducible, it is in fact the algebra of coefficients in irreducible representations, all of which are one dimensional since $\hat{S}$ is abelian. Thus $R(\hat{S})$ is spanned by the group of characters $\hat{\hat{S}}$ of $\hat{S}$. This implies $\mathbb{C}[\hat{\hat{S}}] = R(\hat{S})$.

In the light of Theorem 13.8 one might ask for a direct representation of an invariant strictly positive functional in $\mathbb{C}[S]$ without first resorting to Haar integral on $\hat{S}$. It is simple enough to find it: Let $p:\mathbb{C}[S] \longrightarrow \mathbb{C}$ be given by the projection onto $\mathbb{C}.1$, $1 \in S$. In fact we have the following more general proposition.

15.7. PROPOSITION. *Let* S *be an involutive semigroup with identity and suppose that on the group* H(1) *of units the involution is given by* $s^* = s^{-1}$. *Define a functional* $p:\mathbb{C}[S] \longrightarrow \mathbb{C}$ *so that* $p(\Sigma\, a_s \cdot s) = a_1$. *Then* p *is an invariant positive functional on* $\mathbb{C}[S]$. *It is strictly positive if and only if* S *is a group.*

Proof. Invariance: We will show that the diagram

$$\begin{array}{ccc} \mathbb{C}[S] & \xrightarrow{\quad c \quad} & C[S] \otimes \mathbb{C}[S] \\ {\scriptstyle p}\downarrow & & \downarrow{\scriptstyle p \otimes \mathbb{C}[S]} \\ \mathbb{C} & \longrightarrow & \mathbb{C} \otimes \mathbb{C}[S] \end{array}$$

commutes. Let $s \in S$. Then $c(s) = s \otimes s$ and $(p \otimes 1)c(s) =$ $p(s) \otimes = \begin{cases} 1 \otimes 1_S & \text{if } s = 1_S \\ 0 & \text{otherwise} \end{cases}$. On the other hand, this is exactly the result of applying p and then $\mathbb{C} \longrightarrow \mathbb{C} \otimes \mathbb{C}[S]$ to s. Since the s span $\mathbb{C}[S]$, this proves right invariance. But left invariance is completely analogous.

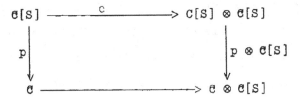

Positivity: Let $a = \Sigma a_s \cdot s$. Then $p(a^*a) = \Sigma \overline{a_s} a_t p(s^*t) = \Sigma\{a_s\overline{a_s}: s \in H(1)\}$ since $p(s^*t) = 0$ if $s^*t \neq 1$, since $s^*t = 1$ implies $s^*, t \in H(1)$ and therefore $s^* = s^{-1}$, and finally since $\overline{a_{s^*}} = a_s$ and $a_{s^{-1}} = \overline{a_s}$. This shows positivity. If $S = H(1)$, then $p(a^*a) = 0$ clearly implies $a = 0$. If $S \neq H(1)$, then there is an $a \in S\backslash H(1)$ such that $p(a^*a) = 0$.

Through 13.9 and the preceding discussion we have established the following theorem.

15.8. THEOREM. *The category of discrete abelian groups is isomorphic to the category of commutative, cocommutative semisimple* Inv-*bigebras with coidentity and an invariant strictly positive functional, and is dual to the category of compact abelian groups. The duality between the categories of discrete and compact abelian groups is given by forming the character groups and giving them the pointwise topology in the discrete case and the uniform topology in the compact one (i.e. the compact open topology in either case).*

This is Pontryagin duality, somewhat enriched by connecting the categories of discrete or compact abelian groups with a suitable category of bigebras. Of course, we could further amplify this theorem by recalling the appropriate category of commutative cocommutative bisimple C*-bigebras with identity.

Much less known than the duality theory of abelian groups is the duality theory of semilattices. A portion of this duality theory was discussed in 14.5.

In the present discussion we start with a discrete semilattice S. We consider it as an involutive semigroup with the identity map of S as involution. By Theorem 15.5.(a), the spectrum $\hat{S}_1$ of $\mathbb{C}[S]$ is isomorphic to $\mathrm{Hom}(S,\mathbb{C}^{\times})$. For each non-zero semicharacter φ of S, the image $\varphi(S)$ is a semilattice in $\mathbb{C}^{\times}$, hence is contained in $\{0,1\}$ (but not in $\{0\}$); thus $\mathrm{Hom}(S,\mathbb{C}^{\times}) = \mathrm{Hom}(S,D)$, whence $\hat{S} = \hat{S}_1$. Now $\hat{S}$ is a compact semilattice with identity, and since it is isomorphic to the subsemilattice $\mathrm{Hom}(S,D) = \mathrm{Hom}(S,\{0,1\})$ of the totally disconnected semilattice $\{0,1\}^S$, it is totally disconnected. By 14.5, $R(\hat{S})$ is a biregular Inv-bigebra which is commutative, cocommutative and idempotent, and which consists of all locally constant functions on $\hat{S}$. The spectrum of $R(\hat{S})$ is $\hat{S}$, i.e. in the terminology of Theorem 15.5 we have $\hat{S}' = \hat{S}$. The semigroup $P(R(\hat{S}))$ of primitive elements contains exactly all semicharacters $\hat{S} \longrightarrow D$, i.e. all morphisms $\hat{S} \longrightarrow \{0,1\}$. Since the non-zero ones must be identity preserving, the semigroup $\hat{\hat{S}}$ of non-zero primitive elements is isomorphic to $\mathrm{Hom}(\hat{S},\{0,1\})$, the semigroup of identity preserving semilattice morphisms. Since the semicharacters of S separate the points of S, the algebra $\mathbb{C}[S]$ may be considered as a subalgebra of $R(\hat{S})$, and S may be considered as a subsemigroup of $\hat{\hat{S}}$ in the obvious fashion: $\hat{s}(f) = f(s)$ for $s \in S$, $f \in \hat{S}$. Since $\mathbb{C}[S]$ separates the points of $\hat{S}$, this algebra is dense in $C(\hat{S})$. We also have Spec $\mathbb{C}[S]$ = $\hat{S}$ = Spec $R(\hat{S})$. If S has an identity, this implies $\mathbb{C}[S] = R(\hat{S})$ and $\hat{\hat{S}} = S$ whenever S is finite. For an arbitrary S let T be a finite subsemilattice of S. There is a commutative diagram

$$\begin{array}{ccc} T & \xrightarrow{\quad i \quad} & S \\ \cap\big\downarrow & & \big\downarrow\cap \\ \mathbb{C}[T] & \xrightarrow{\mathbb{C}[i]} & \mathbb{C}[S] \\ =\big\downarrow & & \big\downarrow\cap \\ R(\hat{T}) & \xrightarrow{R(\hat{i})} & R(\hat{S}) \\ =\big\downarrow & & \big\downarrow\cap \\ C(\hat{T}) & \xrightarrow{C(\hat{i})} & C(\hat{S}) \end{array}$$

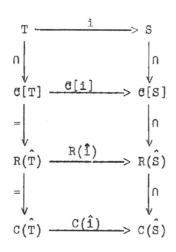

which shows that $\mathbb{C}[i]$ and thus $C(\hat{i})$ are injective. Then $\hat{i}:\hat{S} \longrightarrow \hat{T}$ is surjective. Now suppose $f, g \in \hat{S}$ = Spec $R(\hat{S})$ are such that for all finite subsemilattices $T \subset S$ one has $\hat{i}(f) = \hat{i}(g)$. This means $f|T = g|T$. However, any finite subset of a semilattice generates a finite subsemilattice, since the free semilattice in finitely many generators already is finite. Hence $f|T = g|T$ for all finite subsemilattices T of S means exactly $f = g$. This means that $\hat{S}$ is the projective limit of all of its quotients of the form $\hat{i}:\hat{S} \longrightarrow \hat{T}$. Now suppose that we have a function $\varphi \in R(\hat{S})$. Then (similarly to argument preceding 14.1) there is a finite subsemilattice T such that φ is constant on the sets $\hat{i}^{-1}(\hat{t})$, $\hat{t} \in \hat{T}$. This means that φ is in the image of $R(\hat{T})$ under $R(\hat{i})$. In view of the diagram above we may conclude that φ is in the image of $\mathcal{C}[T]$ under $\mathcal{C}[i]$; in particular, $\varphi \in \mathcal{C}[S]$. Thus $\mathcal{C}[S] = R(\hat{S})$. This then also implies $S = \hat{\hat{S}}$, since the elements of $\hat{\hat{S}}$ are linearly independent in $R(\hat{S})$.

We have in fact proved the following duality theorem:

15.9. THEOREM. *The category of discrete semilattices with identity is isomorphic to the category of biregular commutative cocommutative idempotent bigebras with identity and coidentity over $\mathcal{C}$ and dual to the category of compact totally disconnected semilattices with identity. The duality between the categories of discrete and compact semilattices is given by forming the semilattices of semicharacters and giving them the pointwise topology in the discrete case and the uniform in the compact one (i.e. the compact open topology in either case).*

One should point out that a considerable stock of information is implicit in our discussion which is not at all on the surface. Thus e.g. we have established that the semigroup algebra of any semilattice with identity is biregular, and that any biregular commutative, cocommutative, idempotent bigebra with identity and coidentity is in fact a semilattice algebra and the representation algebra of compact semilattice with identity.

REMARK. In view of earlier results in Section 14, *the category of discrete semilattices with identity is also isomorphic to the category of weakly biregular, commutative, cocommutative, idempotent C*-bigebras with identity and coidentity*.

There are, of course, other categories which are isomorphic to the category of semilattices; in the light of a comment made at the end of Section 14 the category of cocommutative idempotent Boolean bigebras with coidentity is such a category.

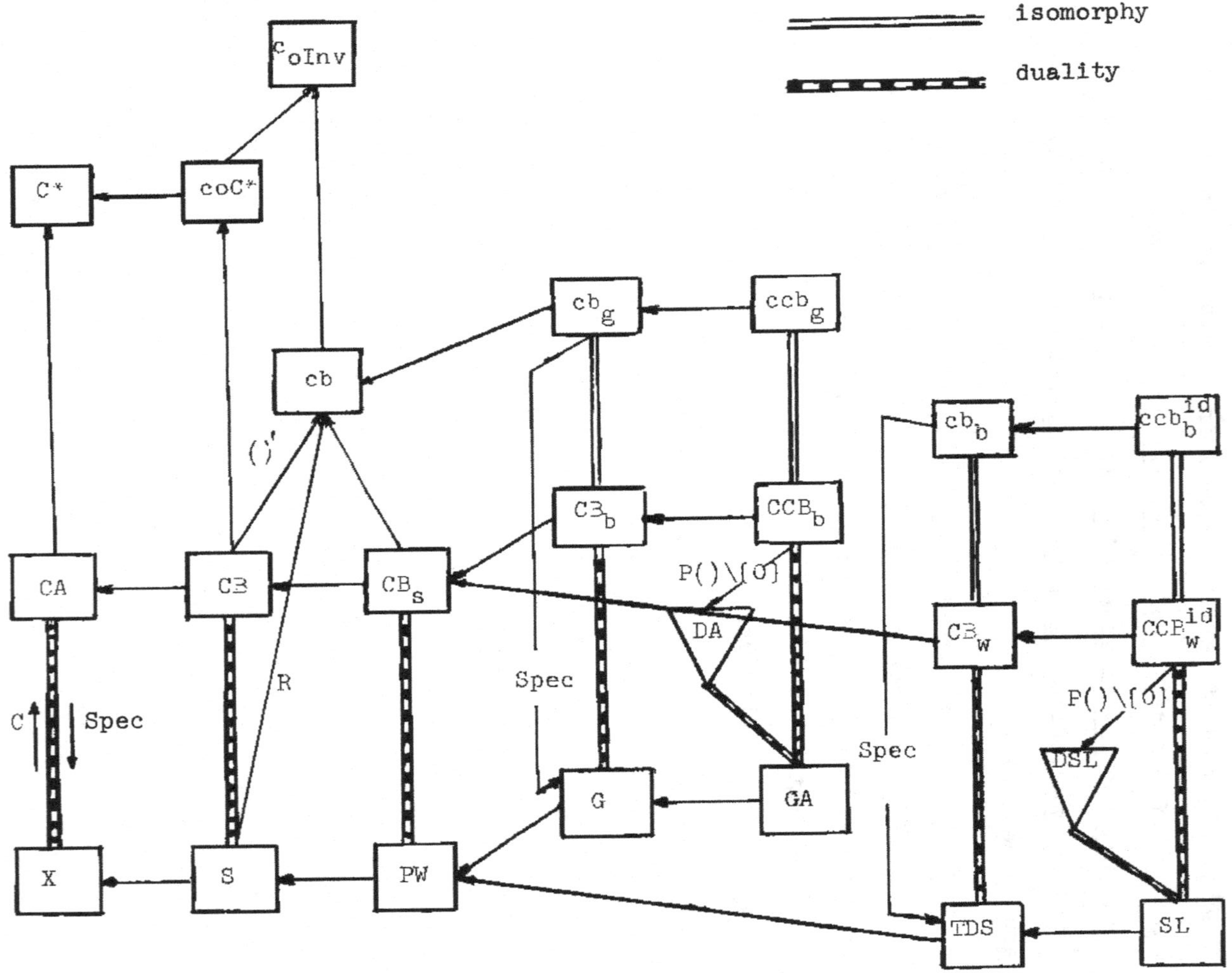
isomorphy
duality
c_{oInv}
C*
coC*
cb
cb_g
ccb_g
()'
CA
CB
CB_s
CB_b
CCB_b
P( )\{0}
DA
R
Spec
C
Spec
Spec
Spec
X
S
PW
G
GA
cb_b
ccb_b^{id}
CB_w
CCB_w^{id}
P( )\{0}
DSL
TDS
SL

EXPLANATION OF THE CHART

All boxes represent categories, all arrows functors or cofunctors. All functors in an (essentially) horizontal direction are inclusion functors. Vertical double lines represent isomorphies and vertical dotted double lines represent dualities. Compositions of dualities and isomorphies yield dualities.

C^* : C^*-algebras with identity
coC^* : C^*-cogebras with identity
coInv: Inv-bigebras with identity
CA : commutative C^*-algebras with identity
CB : commutative C^*-bigebras with identity
CB_s : saturated commutative C^*-bigebras with identity
CB_b : commutative bisimple C^*-bigebras with identity and coidentity
CCB_b : cocommutative objects in CB_b
CB_w : commutative weakly biregular saturated C^*-bigebras with identity
CCB_w^{id}: cocommutative and idempotent objects in CB_w
cb : commutative Inv-bigebras with identity
cb_g : commutative semisimple Inv-bigebras with identity, coidentity and strictly positive invariant functional
ccb_g : cocommutative objects in cb_g
cb_b : commutative, cocommutative, biregular Inv-bigebras with identity and small stalks
ccb_b^{id}: cocommutative and idempotent objects in cb_b
X : compact spaces
S : compact semigroups
PW : compact Peter-Weyl semigroups
G : compact groups
GA : compact abelian groups
DA : discrete abelian groups
TDS : totally disconnected compact Peter-Weyl semigroups
SL : totally disconnected compact semilattices
DSL : discrete semilattices

REFERENCES

[1] BERGLUND, J. and Hofmann, K. H.: Compact Semitopological Semigroups and Weakly Almost Periodic Functions, Lecture Notes in Mathematics. Berlin-Heidelberg-New York: Springer 1967.

[2] CARTIER, P.: Hyperalgèbres et groupes de Lie formels, Sém. Sophus Lie, 2e année, Paris: Secrétariat Mathématique 1957.

[3] DIXMIER, J.: Les algèbres d'operateurs dans l'espace hilbertien, Paris: Gauthier-Villars, 1957.

[4] DIXMIER, J.: Les C*-algèbres et leurs representations, Paris: Gauthier-Villars, 1964.

[5] GROTHENDIECK, A.: Produits tensoriels topologiques et espaces nucléaires, Mem. Amer. Math. Soc. 16, Providence, 1967.

[6] GUICHARDET, A.: Caractères et representations des produits tensoriels de C*-algèbres, Ann. Sci. Ecole Norm. Sup. 81 (1964), 189-206.

[7] GUICHARDET, A.: Tensor Product of C*-algebras, Dokl. Akad. Nauk. SSSR 60 (1965), 986-989 = Soviet Math. Dokl. 6 (1965), 210-213.

[8] HOCHSCHILD, G.: The Structure of Lie Groups, San Francisco-London-Amsterdam: Holden-Day, 1965.

[9] HOFMANN, K. H.: Introduction to the Theory of Compact Groups I. Tulane University Lecture Notes in Mathematics, New Orleans, 1967.

[10] HOFMANN, K. H.: Categories with convergence, exponential functors, and the cohomology of compact abelian groups, Math. Z. 104 (1967), 106-140.

[11] HOFMANN, K. H. and MOSTERT, P. S.: Elements of Compact Semigroups, Columbus (Ohio): Merrill, 1966.

[12] LORENZ, F.: Die Epimorphismen der Ringe von Operatoren, Arch. d. Math. 20 (1969), 48-53.

[13] MITCHELL, B.: Theory of Categories, New York-London: Academic Press, 1965.

[14] POGUNTKE, D: Über kompakte Gruppen, Dissertation Berlin, 1969.

[15] SMITH, H. A.: Positive functionals and representations of tensor products of symmetric Banach algebras, Canadian J. Math. 20 (1968), 1192-1202.

[16] SWEEDLER, M. F.: Integrals for Hopf algebras, Ann. of Math. 89 (1969), 323-335.

[17] WULFSOHN, A.: Produits tensoriels de C*-algebres, Bull. Math. Soc. 87 (1963), 13-27.

[18] WULFSOHN, A.: The primitive spectrum of a tensor product of C*-algebras, Proc. Amer. Math. Soc. 19 (1968), 1094-1096.

Lecture Notes in Mathematics

Bisher erschienen/Already published

Vol. 1: J. Wermer, Seminar über Funktionen-Algebren. IV, 30 Seiten. 1964. DM 3,80 / $ 1.10

Vol. 2: A. Borel, Cohomologie des espaces localement compacts d'après. J. Leray. IV, 93 pages. 1964. DM 9,– / $ 2.60

Vol. 3: J. F. Adams, Stable Homotopy Theory. Third edition. IV, 78 pages. 1969. DM 8,– / $ 2.20

Vol. 4: M. Arkowitz and C. R. Curjel, Groups of Homotopy Classes. 2nd. revised edition. IV, 36 pages. 1967. DM 4,80 / $ 1.40

Vol. 5: J.-P. Serre, Cohomologie Galoisienne. Troisième édition. VIII, 214 pages. 1965. DM 18,– / $ 5.00

Vol. 6: H. Hermes, Eine Termlogik mit Auswahloperator. IV, 42 Seiten. 1965. DM 5,80 / $ 1.60

Vol. 7: Ph. Tondeur, Introduction to Lie Groups and Transformation Groups. Second edition. VIII, 176 pages. 1969. DM 14,– / $ 3.80

Vol. 8: G. Fichera, Linear Elliptic Differential Systems and Eigenvalue Problems. IV, 176 pages. 1965. DM 13,50 / $ 3.80

Vol. 9: P. L. Ivănescu, Pseudo-Boolean Programming and Applications. IV, 50 pages. 1965. DM 4,80 / $ 1.40

Vol. 10: H. Lüneburg, Die Suzukigruppen und ihre Geometrien. VI, 111 Seiten. 1965. DM 8,– / $ 2.20

Vol. 11: J.-P. Serre, Algèbre Locale. Multiplicités. Rédigé par P. Gabriel. Seconde édition. VIII, 192 pages. 1965. DM 12,– / $ 3.30

Vol. 12: A. Dold, Halbexakte Homotopiefunktoren. II, 157 Seiten. 1966. DM 12,– / $ 3.30

Vol. 13: E. Thomas, Seminar on Fiber Spaces. IV, 45 pages. 1966. DM 4,80 / $ 1.40

Vol. 14: H. Werner, Vorlesung über Approximationstheorie. IV, 184 Seiten und 12 Seiten Anhang. 1966. DM 14,– / $ 3.90

Vol. 15: F. Oort, Commutative Group Schemes. VI, 133 pages. 1966. DM 9,80 / $ 2.70

Vol. 16: J. Pfanzagl and W. Pierlo, Compact Systems of Sets. IV, 48 pages. 1966. DM 5,80 / $ 1.60

Vol. 17: C. Müller, Spherical Harmonics. IV, 46 pages. 1966. DM 5,– / $ 1.40

Vol. 18: H.-B. Brinkmann und D. Puppe, Kategorien und Funktoren. XII, 107 Seiten, 1966. DM 8,– / $ 2.20

Vol. 19: G. Stolzenberg, Volumes, Limits and Extensions of Analytic Varieties. IV, 45 pages. 1966. DM 5,40 / $ 1.50

Vol. 20: R. Hartshorne, Residues and Duality. VIII, 423 pages. 1966. DM 20,– / $ 5.50

Vol. 21: Seminar on Complex Multiplication. By A. Borel, S. Chowla, C. S. Herz, K. Iwasawa, J.-P. Serre. IV, 102 pages. 1966. DM 8,– / $ 2.20

Vol. 22: H. Bauer, Harmonische Räume und ihre Potentialtheorie. IV, 175 Seiten. 1966. DM 14,– / $ 3.90

Vol. 23: P. L. Ivănescu and S. Rudeanu, Pseudo-Boolean Methods for Bivalent Programming. 120 pages. 1966. DM 10,– / $ 2.80

Vol. 24: J. Lambek, Completions of Categories. IV, 69 pages. 1966. DM 6,80 / $ 1.90

Vol. 25: R. Narasimhan, Introduction to the Theory of Analytic Spaces. IV, 143 pages. 1966. DM 10,– / $ 2.80

Vol. 26: P.-A. Meyer, Processus de Markov. IV, 190 pages. 1967. DM 15,– / $ 4.20

Vol. 27: H. P. Künzi und S. T. Tan, Lineare Optimierung großer Systeme. VI, 121 Seiten. 1966. DM 12,– / $ 3.30

Vol. 28: P. E. Conner and E. E. Floyd, The Relation of Cobordism to K-Theories. VIII, 112 pages. 1966. DM 9,80 / $ 2.70

Vol. 29: K. Chandrasekharan, Einführung in die Analytische Zahlentheorie. VI, 199 Seiten. 1966. DM 16,80 / $ 4.70

Vol. 30: A. Frölicher and W. Bucher, Calculus in Vector Spaces without Norm. X, 146 pages. 1966. DM 12,– / $ 3.30

Vol. 31: Symposium on Probability Methods in Analysis. Chairman. D. A. Kappos. IV, 329 pages. 1967. DM 20,– / $ 5.50

Vol. 32: M. André, Méthode Simpliciale en Algèbre Homologique et Algèbre Commutative. IV, 122 pages. 1967. DM 12,– / $ 3.30

Vol. 33: G. I. Targonski, Seminar on Functional Operators and Equations. IV, 110 pages. 1967. DM 10,– / $ 2.80

Vol. 34: G. E. Bredon, Equivariant Cohomology Theories. VI, 64 pages. 1967. DM 6,80 / $ 1.90

Vol. 35: N. P. Bhatia and G. P. Szegö, Dynamical Systems. Stability Theory and Applications. VI, 416 pages. 1967. DM 24,– / $ 6.60

Vol. 36: A. Borel, Topics in the Homology Theory of Fibre Bundles. VI, 95 pages. 1967. DM 9,– / $ 2.50

Vol. 37: R. B. Jensen, Modelle der Mengenlehre. X, 176 Seiten. 1967. DM 14,– / $ 3.90

Vol. 38: R. Berger, R. Kiehl, E. Kunz und H.-J. Nastold, Differentialrechnung in der analytischen Geometrie IV, 134 Seiten. 1967 DM 12,– / $ 3.30

Vol. 39: Séminaire de Probabilités I. II, 189 pages. 1967. DM 14,– / $ 3.90

Vol. 40: J. Tits, Tabellen zu den einfachen Lie Gruppen und ihren Darstellungen. VI, 53 Seiten. 1967. DM 6,80 / $ 1.90

Vol. 41: A. Grothendieck, Local Cohomology. VI, 106 pages. 1967. DM 10,– / $ 2.80

Vol. 42: J. F. Berglund and K. H. Hofmann, Compact Semitopological Semigroups and Weakly Almost Periodic Functions. VI, 160 pages. 1967. DM 12,– / $ 3.30

Vol. 43: D. G. Quillen, Homotopical Algebra. VI, 157 pages. 1967. DM 14,– / $ 3.90

Vol. 44: K. Urbanik, Lectures on Prediction Theory. IV, 50 pages. 1967. DM 5,80 / $ 1.60

Vol. 45: A. Wilansky, Topics in Functional Analysis. VI, 102 pages. 1967. DM 9,60 / $ 2.70

Vol. 46: P. E. Conner, Seminar on Periodic Maps. IV, 116 pages. 1967. DM 10,60 / $ 3.00

Vol. 47: Reports of the Midwest Category Seminar I. IV, 181 pages. 1967. DM 14,80 / $ 4.10

Vol. 48: G. de Rham, S. Maumary et M. A. Kervaire, Torsion et Type Simple d'Homotopie. IV, 101 pages. 1967. DM 9,60 / $ 2.70

Vol. 49: C. Faith, Lectures on Injective Modules and Quotient Rings. XVI, 140 pages. 1967. DM 12,80 / $ 3.60

Vol. 50: L. Zalcman, Analytic Capacity and Rational Approximation. VI, 155 pages. 1968. DM 13.20 / $ 3.70

Vol. 51: Séminaire de Probabilités II. IV, 199 pages. 1968. DM 14,– / $ 3.90

Vol. 52: D. J. Simms, Lie Groups and Quantum Mechanics. IV, 90 pages. 1968. DM 8,– / $ 2.20

Vol. 53: J. Cerf, Sur les difféomorphismes de la sphère de dimension trois (Γ_4 = O). XII, 133 pages. 1968. DM 12,– / $ 3.30

Vol. 54: G. Shimura, Automorphic Functions and Number Theory. VI, 69 pages. 1968. DM 8,– / $ 2.20

Vol. 55: D. Gromoll, W. Klingenberg und W. Meyer, Riemannsche Geometrie im Großen. VI, 287 Seiten. 1968. DM 20,– / $ 5.50

Vol. 56: K. Floret und J. Wloka, Einführung in die Theorie der lokalkonvexen Räume. VIII, 194 Seiten. 1968. DM 16,– / $ 4.40

Vol. 57: F. Hirzebruch und K. H. Mayer, O (n)-Mannigfaltigkeiten, exotische Sphären und Singularitäten. IV, 132 Seiten. 1968. DM 10,80 / $ 3.00

Vol. 58: Kuramochi Boundaries of Riemann Surfaces. IV, 102 pages. 1968. DM 9,60 / $ 2.70

Vol. 59: K. Jänich, Differenzierbare G-Mannigfaltigkeiten. VI, 89 Seiten. 1968. DM 8,– / $ 2.20

Vol. 60: Seminar on Differential Equations and Dynamical Systems. Edited by G. S. Jones. VI, 106 pages. 1968. DM 9,60 / $ 2.70

Vol. 61: Reports of the Midwest Category Seminar II. IV, 91 pages. 1968. DM 9,60 / $ 2.70

Vol. 62: Harish-Chandra, Automorphic Forms on Semisimple Lie Groups X, 138 pages. 1968. DM 14,– / $ 3.90

Vol. 63: F. Albrecht, Topics in Control Theory. IV, 65 pages. 1968. DM 6,80 / $ 1.90

Vol. 64: H. Berens, Interpolationsmethoden zur Behandlung von Approximationsprozessen auf Banachräumen. VI, 90 Seiten. 1968. DM 8,– / $ 2.20

Vol. 65: D. Kölzow, Differentiation von Maßen. XII, 102 Seiten. 1968. DM 8,– / $ 2.20

Vol. 66: D. Ferus, Totale Absolutkrümmung in Differentialgeometrie und -topologie. VI, 85 Seiten. 1968. DM 8,– / $ 2.20

Vol. 67: F. Kamber and P. Tondeur, Flat Manifolds. IV, 53 pages. 1968. DM 5,80 / $ 1.60

Vol. 68: N. Boboc et P. Mustață, Espaces harmoniques associés aux opérateurs différentiels linéaires du second ordre de type elliptique. VI, 95 pages. 1968. DM 8,60 / $ 2.40

Vol. 69: Seminar über Potentialtheorie. Herausgegeben von H. Bauer. VI, 180 Seiten. 1968. DM 14,80 / $ 4.10

Vol. 70: Proceedings of the Summer School in Logic. Edited by M. H. Löb. IV, 331 pages. 1968. DM 20,– / $ 5.50

Vol. 71: Séminaire Pierre Lelong (Analyse), Année 1967 – 1968. VI, 19 pages. 1968. DM 14,– / $ 3.90

Bitte wenden / Continued

Vol. 72: The Syntax and Semantics of Infinitary Languages. Edited by J. Barwise. IV, 268 pages. 1968. DM 18,– / $ 5.00

Vol. 73: P. E. Conner, Lectures on the Action of a Finite Group. IV, 123 pages. 1968. DM 10,– / $ 2.80

Vol. 74: A. Fröhlich, Formal Groups. IV, 140 pages. 1968. DM 12,– / $ 3.30

Vol. 75: G. Lumer, Algèbres de fonctions et espaces de Hardy. VI, 80 pages. 1968. DM 8,– / $ 2.20

Vol. 76: R. G. Swan, Algebraic K-Theory. IV, 262 pages. 1968. DM 18,– / $ 5.00

Vol. 77: P.-A. Meyer, Processus de Markov: la frontière de Martin. IV, 123 pages. 1968. DM 10,– / $ 2.80

Vol. 78: H. Herrlich, Topologische Reflexionen und Coreflexionen. XVI, 166 Seiten. 1968. DM 12,– / $ 3.30

Vol. 79: A. Grothendieck, Catégories Cofibrées Additives et Complexe Cotangent Relatif. IV, 167 pages. 1968. DM 12,– / $ 3.30

Vol. 80: Seminar on Triples and Categorical Homology Theory. Edited by B. Eckmann. IV, 398 pages. 1969. DM 20,– / $ 5.50

Vol. 81: J.-P. Eckmann et M. Guenin, Méthodes Algébriques en Mécanique Statistique. VI, 131 pages. 1969. DM 12,– / $ 3.30

Vol. 82: J. Wloka, Grundräume und verallgemeinerte Funktionen. VIII, 131 Seiten. 1969. DM 12,– / $ 3.30

Vol. 83: O. Zariski, An Introduction to the Theory of Algebraic Surfaces. IV, 100 pages. 1969. DM 8,– / $ 2.20

Vol. 84: H. Lüneburg, Transitive Erweiterungen endlicher Permutationsgruppen. IV, 119 Seiten. 1969. DM 10,– / $ 2.80

Vol. 85: P. Cartier et D. Foata, Problèmes combinatoires de commutation et réarrangements. IV, 88 pages. 1969. DM 8,– / $ 2.20

Vol. 86: Category Theory, Homology Theory and their Applications I. Edited by P. Hilton. VI, 216 pages. 1969. DM 16,– / $ 4.40

Vol. 87: M. Tierney, Categorical Constructions in Stable Homotopy Theory. IV, 65 pages. 1969. DM 6,– / $ 1.70

Vol. 88: Séminaire de Probabilités III. IV, 229 pages. 1969. DM 18,– / $ 5.00

Vol. 89: Probability and Information Theory. Edited by M. Behara, K. Krickeberg and J. Wolfowitz. IV, 256 pages. 1969. DM 18,– / $ 5.00

Vol. 90: N. P. Bhatia and O. Hajek, Local Semi-Dynamical Systems. II, 157 pages. 1969. DM 14,– / $ 3.90

Vol. 91: N. N. Janenko, Die Zwischenschrittmethode zur Lösung mehrdimensionaler Probleme der mathematischen Physik. VIII, 194 Seiten. 1969. DM 16,80 / $ 4.70

Vol. 92: Category Theory, Homology Theory and their Applications II. Edited by P. Hilton. V, 308 pages. 1969. DM 20,– / $ 5.50

Vol. 93: K. R. Parthasarathy, Multipliers on Locally Compact Groups. III, 54 pages. 1969. DM 5,60 / $ 1.60

Vol. 94: M. Machover and J. Hirschfeld, Lectures on Non-Standard Analysis. VI, 79 pages. 1969. DM 6,– / $ 1.70

Vol. 95: A. S. Troelstra, Principles of Intuitionism. II, 111 pages. 1969. DM 10,– / $ 2.80

Vol. 96: H.-B. Brinkmann und D. Puppe, Abelsche und exakte Kategorien, Korrespondenzen. V, 141 Seiten. 1969. DM 10,– / $ 2.80

Vol. 97: S. O. Chase and M. E. Sweedler, Hopf Algebras and Galois theory. II, 133 pages. 1969. DM 10,– / $ 2.80

Vol. 98: M. Heins, Hardy Classes on Riemann Surfaces. III, 106 pages. 1969. DM 10,– / $ 2.80

Vol. 99: Category Theory, Homology Theory and their Applications III. Edited by P. Hilton. IV, 489 pages. 1969. DM 24,– / $ 6.60

Vol. 100: M. Artin and B. Mazur, Etale Homotopy. II, 196 Seiten. 1969. DM 12,– / $ 3.30

Vol. 101: G. P. Szegö et G. Treccani, Semigruppi di Trasformazioni Multivoche. VI, 177 pages. 1969. DM 14,– / $ 3.90

Vol. 102: F. Stummel, Rand- und Eigenwertaufgaben in Sobolewschen Räumen. VIII, 386 Seiten. 1969. DM 20,– / $ 5.50

Vol. 103: Lectures in Modern Analysis and Applications I. Edited by C. T. Taam. VII, 162 pages. 1969. DM 12,– / $ 3.30

Vol. 104: G. H. Pimbley, Jr., Eigenfunction Branches of Nonlinear Operators and their Bifurcations. II, 128 pages. 1969. DM 10,– / $ 2.80

Vol. 105: R. Larsen, The Multiplier Problem. VII, 284 pages. 1969. DM 18,– / $ 5.00

Vol. 106: Reports of the Midwest Category Seminar III. Edited by S. Mac Lane. III, 247 pages. 1969. DM 16,– / $ 4.40

Vol. 107: A. Peyerimhoff, Lectures on Summability. III, 111 pages. 1969. DM 8,– / $ 2.20

Vol. 108: Algebraic K-Theory and its Geometric Applications. Edited by R. M.F. Moss and C. B. Thomas. IV, 86 pages. 1969. DM 6,– / $ 1.70

Vol. 109: Conference on the Numerical Solution of Differential Equations. Edited by J. Ll. Morris. VI, 275 pages. 1969. DM 18,– / $ 5.00

Vol. 110: The Many Facets of Graph Theory. Edited by G. Chartrand and S. F. Kapoor. VIII, 290 pages. 1969. DM 18,– / $ 5.00

Vol. 111: K. H. Mayer, Relationen zwischen charakteristischen Zahlen. III, 99 Seiten. 1969. DM 8,– / $ 2.20

Vol. 112: Colloquium on Methods of Optimization. Edited by N. N. Moiseev. IV, 293 pages. 1970. DM 18,– / $ 5.00

Vol. 113: R. Wille, Kongruenzklassengeometrien. III, 99 Seiten. 1970. DM 8,– / $ 2.20

Vol. 114: H. Jacquet and R. P. Langlands, Automorphic Forms on GL (2). VII, 548 pages. 1970. DM 24,– / $ 6.60

Vol. 115: K. H. Roggenkamp and V. Huber-Dyson, Lattices over Orders I. XIX, 290 pages. 1970. DM 18,– / $ 5.00

Vol. 116: Séminaire Pierre Lelong (Analyse) Année 1969. IV, 195 pages. 1970. DM 14,– / $ 3.90

Vol. 117: Y. Meyer, Nombres de Pisot, Nombres de Salem et Analyse Harmonique. 63 pages. 1970. DM 6,– / $ 1.70

Vol. 118: Proceedings of the 15th Scandinavian Congress, Oslo 1968. Edited by K. E. Aubert and W. Ljunggren. IV, 162 pages. 1970. DM 12,– / $ 3.30

Vol. 119: M. Raynaud, Faisceaux amples sur les schémas en groupes et les espaces homogènes. III, 219 pages. 1970. DM 14,– / $ 3.90

Vol. 120: D. Siefkes, Büchi's Monadic Second Order Successor Arithmetic. XII, 130 Seiten. 1970. DM 12,– / $ 3.30

Vol. 121: H. S. Bear, Lectures on Gleason Parts. III, 47 pages. 1970. DM 6,– / $ 1.70

Vol. 122: H. Zieschang, E. Vogt und H.-D. Coldewey, Flächen und ebene diskontinuierliche Gruppen. VIII, 203 Seiten. 1970. DM 16,– / $ 4.40

Vol. 123: A. V. Jategaonkar, Left Principal Ideal Rings. VI, 145 pages. 1970. DM 12,– / $ 3.30

Vol. 124: Séminare de Probabilités IV. Edited by P. A. Meyer. IV, 282 pages. 1970. DM 20,– / $ 5.50

Vol. 125: Symposium on Automatic Demonstration. V, 310 pages. 1970. DM 20,– / $ 5.50

Vol. 126: P. Schapira, Théorie des Hyperfonctions. XI, 157 pages. 1970. DM 14,– / $ 3.90

Vol. 127: I. Stewart, Lie Algebras. IV, 97 pages. 1970. DM 10,– / $ 2.80

Vol. 128: M. Takesaki, Tomita's Theory of Modular Hilbert Algebras and its Applications. II, 123 pages. 1970. DM 10,– / $ 2.80

Vol. 129: K. H. Hofmann, The Duality of Compact Semigroups and C*- Bigebras. XII, 142 pages. 1970. DM 14,– / $ 3.90